Advance Praise for The Ice Chronicles—

"Almost everybody seems to be aware of the fact that significant changes have been taking place in our climate—but . . . *The Ice Chronicles* tells us, in marvelous detail, exactly why these changes are taking place, and the fact that this earth-warming process has been going on steadily for many centuries . . . This book is a remarkable document. Everyone should read it and heed its warning."

—BRADFORD WASHBURN,
Founding Director,
Boston Museum of Science

"Paul Mayewski and Frank White have written an exciting account of what has been learned about past climates from drilling into ice sheets. From scientists to lay persons, readers will be involved on a first-hand basis in one of the greatest scientific activities of our time, understanding how climate changes and the impact of these changes on human society. It's simply a great read!"

—W. BERRY LYONS,
Byrd Polar Research Center,
Ohio State University

"*The Ice Chronicles* is at once a thoroughly enjoyable adventure story of intrepid explorers, an exciting history of expeditions into alpine and polar regions, an introduction to glaciology, and a chronicle of our sometimes turbulent climate. Ice layers provide a detailed script of Rapid Climate Change Events, or RCCEs, that have punctuated periods of relative stability—often with profound impacts on the evolution of species and societies. *The Ice Chronicles* reveals how multiple human activities—changing land use/land cover, stratospheric ozone and greenhouse gas concentrations—are together altering natural climate variability and destabilizing the climate system. This work demonstrates that we have underestimated the potential for abrupt climate change, and the authors do not flinch from asserting the need for bold solutions."

—PAUL R. EPSTEIN, M.D., M.P.H.,
Associate Director,
Center for Health and the Global Environment,
Harvard Medical School

"In *The Ice Chronicles*, Professor Mayewski shares vividly with us not only the exciting science of the Earth but also the extraordinary adventure of doing this science. Shifting glaciers, expansive ice sheets, towering mountaintops, and enormous valleys fill this remarkable canvas, and through them we are able to peer back into time and thereby gaze into the future. This remarkable book should be read by all citizens of the planet Earth."

—BERRIEN MOORE,
Institute for the Study of Earth, Oceans, & Space,
University of New Hampshire

"Through this book you will travel over the coldest places on Earth, glaciers, Greenland, Antarctica, and high mountains in the Himalayas. You will participate in the birth of an ambitious deep drilling project and discover the wealth of ice archives . . . Beyond the fascination of exploration and scientific discoveries, you will be disturbed by the expected impact of global warming on our societies. This is a challenge that all of us have to face at the dawn of the twenty-first century."

—CLAUDE LORIUS,
Laboratoire de Glaciologie et
Géophysique de l'Environnement, France

"Locked within ice cores are secrets about Earth's past—its geology, its climate, and its ecosystems. Leading expeditions to the most inhospitable corners of the globe, Mayewski and co-workers have unlocked many of these secrets to assist the current debate on climate change. This book portrays the dangers of these expeditions while illustrating the excitement of scientific discovery and the urgency of incorporating science into policy-making."

—CLIFF I. DAVIDSON,
Editor, *Clean Hands: Clair Patterson's Crusade
Against Environmental Lead Contamination*

"The book is told from the perspective of Paul Mayewski, but the reader is also introduced to many of the scientists behind the great project of coring through the 3,000-meter-thick ice sheet of central Greenland. Perhaps this is the single project that has added the most to our knowledge of climate history. The authors give details about the long, cold, often monotonous field and laboratory work, and insight into the force that drives scientists to spend many years of their lives in some of the most remote and forbidding places on Earth. Through it all, Mayewski and White also manage to describe the beauty of the arctic and alpine regions, and the very special experience of working in a group isolated by distance and weather."

—WIBJÖRN KARLÉN, Professor,
Department of Physical Geography and Quaternary Geology,
Stockholm University

The Ice Chronicles

THE ICE CHRONICLES

The Quest to Understand Global Climate Change

Paul Andrew Mayewski

&

Frank White

UNIVERSITY OF NEW HAMPSHIRE

PUBLISHED BY

UNIVERSITY PRESS OF NEW ENGLAND

HANOVER AND LONDON

University of New Hampshire

Published by University Press of New England, Hanover, NH 03755

© 2002 by Paul Andrew Mayewski and Frank White

Printed in the United States of America

5 4 3 2

Watercolor by Lyn Weisman Mayewski entitled "Ascent of the Nun Kun Plateau, Ladakh, India, in Search of the Ice Chronicles."

Library of Congress Cataloging-in-Publication Data

Mayewski, Paul A.
   The ice chronicles / by Paul Andrew Mayewski and Frank White.
     p. cm.
Includes bibliographical references (p.   ).
  ISBN 1–58465–061–3 (cloth : alk. paper)
  1. Glaciers. 2. Climatic changes. I. White, Frank, 1944–   II. Title.
  GB2405 .M28 2002
  551.6—dc21   2001004980

From Paul to Lyn

&

From Frank to Donna

Contents

Illustrations

Foreword

Little seems more sure than the obvious fact that spring follows winter and summer, spring. Those in the sunbelt may welcome the relatively seasonless climate of the South, whereas those of us enamored of deep and quiet snow, like author Paul Mayewski, find fascination in the frigid wilds of Antarctica's landscape. Here, from firsthand experience with one of Nature's longest and greatest books—the record of weather preserved deep in ice—Mayewski tells us as true a story as can be gleaned about the realities of climate stability and patterns of weather. In *The Ice Chronicles,* we are gifted with a genuinely popular account of weather shifts, one in which casual opinion and "the misplaced concreteness" of theory are eclipsed by prodigious quantities of information. The clear voice of co-author Frank White, an expert on the experience of cosmonauts and astronauts who circled or escaped our Earth, penetrates the planet's cloud shroud. Finally, we follow here a guide map to climate change that dispenses with hype and hyperbole, the stock-in-trade of this tricky, international, and politically charged issue

Ask yourself now: Have you ever lived for more than a few months in a climate that genuinely differs from that of your childhood? Those of us who have know in some deep place in our being that weather and climate powerfully alter our moods, motivations, and habits. Some take for granted that more or less what came last year will come again this time. Few in the northeastern megalopolis of North America know that just after the great explosion of Krakatoa in 1883 the summer in Vermont was obliterated. Snow remained on the ground in July, and no harvest followed. None of us, even scientists with professional interests in the history of planet Earth and its biosphere, fully knows the measured story of the great ice core that White and Mayewski chronicle.

As much as any of us, Mayewski enjoys the balmy summer sun and

warm tropical breezes, but unlike nearly everyone, he has spent his professional life in Greenland, Antarctica, and other inhospitable land-
scapes, where he has worked as a geologist and climatologist in pursuit
of the rhythm and reason for staggering winds and bone-chilling cold.
"Dancing snow" means trouble. He describes a time in Antarctica when
it gave him no option and tells how his battle and that of his colleagues
required patience: "We stopped right there, pitched our tents, and prepared to wait the storm out. . . . We had brought along a 120-pound generator that kept jumping around on its platform, and we needed every
piece of climbing rope and all of our extra climbing screws to tie down
the generator and our tent." They knew that such storms might take up
to ten days to dissipate. "Rescue was unthinkable; we couldn't expect the
people there at McMurdo Station [100 miles away] to risk their own
lives. . . . We took turns holding up a piece of plywood against the wind
wall of the tent to keep the whole thing from collapsing. That fabric
was all that stood between us and a very serious situation." They
couldn't even risk sleeping. Only Lady Luck limited that particular
tempest to 72 rather than 240 hours and granted them a faster escape.

People respond differently to rapid environmental changes, but all of
us respond. Mayewski remembers a Himalayan trip during which he
awoke to the sound of some porters who had returned to camp after an
evening down-mountain with their families. Unbelievably, they had de-
parted at dusk from 18,000 feet, eaten well only after sundown (for it
was Ramadan), slept a few hours at 7,000 feet, began their return at 2:30
A.M., and now, with their beagle in tow, refreshed but *still barefoot,* they
were ready for and cheerful about the next leg of the climb. This, for
these porters, was the way of life. But why was it becoming so for a
young professor from New Hampshire? Curiosity about why climates
are stable and why they change lies at the heart of why they wrote and
why we read this story.

White's way with words and his own fascination with Mayewski's
story are overshadowed only by the observational and analytical science of the story chronicled here. The real central characters are not
the co-authors, the Himalayan porters, the bush pilots and international scientific staff at Antarctica, the National Science Foundation
sages who recognized the importance of the work and funded it, the
twenty-five institutions and their indefatigable researchers who examine the revelatory foraminiferan shells in the sediment, or the political
activists who rail against public complacency and private greed. No, the

real hero is a long, skinny chunk of icy layered sediment that faithfully records 40,000 human lifetimes of weather change.

How did the world obtain this missive from the past? How do neatly ordered layers of rubble chronicle climate history? No argument based on human documentation, either oral or written, has any validity at all in this contentious, newsy issue. Rather, an incontrovertible, if mute, authority has already recorded vast swings in the world's weather patterns. The rock record, the heroic core with its oldest layer at the bottom and its youngest at the top, graphically illustrates the geologist's "law of superposition."

We complete the Mayewski–White adventure story with a deep understanding of this rocky biopsy of the Earth's skin. When the rotating drill bit of the long core finally struck bedrock and the drilling ceased, what would turn out to be a sample of the world's weather every 2 years for the last 10,000 years (and every 15 years for the period from 10,000 to 45,000 years and every 50 years from 45,000 to 110,000 years) was in hand. On that day in 1993 when the bit hit bottom, the scientists raised plastic champagne glasses in the wind. They had "witnessed an historical event." The story of how the great ice chronicle and its shorter brethren were acquired and forced to yield their scientific secrets, as round and shocking and funny and sad as any Icelandic saga, is told with uncharacteristic dramatic intensity. The long cylinder, divvied up for analysis and subject to enormous powers of investigation, was worth all the effort.

Now the ancient climate record has become a single tale. Far more compelling than any storyboard of the imagination, we have a true history of environmental change, starts and stops, trends, and imperatives at annual, decadal, and century time frames for a total of more than one hundred millennia. Climate—as changes in the seasons and their average properties—comes alive before our eyes. Mayewski and White explain how change at every level, day to month to century to millennium to eon, has been the rule. In accessible language, they squeeze the overblown literature to yield its concentrated juice.

The forces of change spur these authors to offer ten reasonable "principles for a positive future." Of course, the list is naive and oversimplified, but naive, optimistic, adventurous, and curious are species characteristics that Mayewski and White share with most of our fellow *Homo sapiens*. They show how climate change and its preoccupations have been linked to the rise and fall of agricultural regimes, civilizations, businesses, and property values from the beginning of recorded history. Fellow scientists and their funding administrators, journalists, and explorers of today, all enlightened by the excesses and hardships of the peopled environments of the past, march through these pages. Those who collect and analyze the great ice core lack the overview: They are always too busy, too specialized, or too uninformed to comprehend the entire 3,060-meter-long record. Indeed, every one of us has been too preoccupied or too underappreciative to perform the public service of describing the vagaries and commitments of our past environments. The long, skinny, truthful core from which climate change prognostications are ultimately fabricated has lain buried beneath the ice. That is, until now.

Spring 2001   Lynn Margulis

Preface

We met when Frank came to the Institute for the Study of Earth, Oceans, and Space (EOS) at the University of New Hampshire in 1994 to do a writing project for the Institute. Paul was the first person to be interviewed for the project, and halfway through the discussion of drilling into glaciers, climbing mountains with barefoot porters carrying beagles, and falling into crevasses, Frank said, "Paul, you have a great book here. You really ought to write it." A letter from Paul to Frank after that first meeting led to a collaboration that has lasted for more than six years and survived trips to Antarctica and other far-flung corners of the world, family illnesses and death, job changes, and many rewrites.

We've been able to persevere through all these obstacles because of a mutual personal respect that has grown as we've come to know each other better and a shared belief that this is a fascinating story that needs to be told. It's not just the personal saga of Paul Mayewski, although his adventures are fascinating, but also the description of science as it is carried out today and the unfolding drama of changes in the Earth over the millennia. It is also a cautionary tale, providing some important insights into the impact of climate on civilization and the effects it may have on everyone in the near future.

Although our lives are very different, and our interests seem opposite, we share a strong interest in exploration and its value for the human spirit. Frank's work as an author about the impact of space exploration on human consciousness has provided a valuable foundation for sharing in the account of Paul's work on the impact of climate change on our small planet.

We hope our readers will enjoy the adventure of *The Ice Chronicles*, learn from the science presented here, and use the information in this book to become better informed citizens at a time when important decisions will be made around the world about climate change in general and global warming and quality of our air and water in particular.

Since we finished the book in the fall of 2000, the following events have occurred:

• President Bush withdrew from the dialogue on the Kyoto Treaty, to the consternation of other nations that had been involved in trying to reach some consensus on climate change. In a State Department memo on the subject, the Bush administration said that it was rejecting the treaty "under any circumstances" ("U.S. Pessimistic on Warming Treaty," John Heilprin, The Associated Press, April 20, 2001).

• The Intergovernmental Panel on Climate Change (IPCC), in its 2000 report, went far beyond its 1995 report in suggesting that global warming is a reality.

• President Bush retreated on a campaign promise to restrict CO_2 emissions from power plants in the United States, to the chagrin of numerous environmental groups and ordinary citizens in this country.

• *Time* magazine featured global warming as its cover story.

• The National Council for Science and the Environment released a report based on the deliberations of more than 450 scientists, policymakers, and others calling for the creation of a new interdisciplinary "science of sustainability" ("Report of 1st National Conference on Science, Policy & the Environment Released: Hill Briefings Scheduled" from Kevin Hutton, webmaster, National Council, April 23, 2001).

• The National Academy of Sciences, in a report requested by President Bush, confirmed the reality of global warming.

• Organizations in Europe began a boycott of companies seen as contributing to global warming.

• President Bush announced new policies on global warming.

• Terrorist attacks on the United States moved the issues of climate change and global warming to the background of public awareness.

It is understandable that environmental issues will not be at the forefront of the policy debate for some time because of a transformed political situation worldwide. However, the problems and challenges discussed in this book will continue to be important, and will eventually demand the attention of the global community once again.

We could cite many other examples that illustrate the growing interest in and concern about climate change in general and global warming in particular. If we attempted to incorporate all these recent developments into *The Ice Chronicles*, the book would simply never get published. For that reason, we urge readers to see the book not as an up-to-date report on current events in the field but rather as a framework for understanding those events. We hope that, after reading *The Ice Chronicles,* you will have a more informed opinion about President Bush's decisions or the IPCC report. Just as the ice cores retrieved from glaciers around the world provide a backdrop for understanding current climate, so does this book provide a context for understanding current policy decisions.

(A note on "voice": The subject of this book is primarily Paul's research as a glaciologist. For that reason, it is written in his voice, and whenever we use the personal pronoun "I," the reference is to Paul. At the same time, the book is truly a collaboration, and that is why we share co-authorship.)

P. M. and F. W.

Acknowledgments

PAUL: I want to begin my acknowledgments by thanking my wife, Lyn, who has shared so much of my life. Without her, it would not have been possible for me to have undertaken the expeditions and science described in this book. She gives me the freedom to live my dreams and most importantly the love that has supported me in both the good and bad times we all face in our lives. Together with B. J., Kruthers, Mac, Maggie, Bob, and Tiana, we have been very fortunate.

Support to follow my own path came from day one in my life. My mother, Lisa Asnis, gave me her gift for optimism, a sense of humor, and her will to achieve a goal. My father, Paul Mayewski, gave me the desire to explore, the strength to persevere, and a sense of humor in the face of adversity. My stepmother, Maxine S. Mayewski, gave me an understanding of higher education. My stepfather, Al Asnis, taught me how to relax. My parents-in-law, Mike and Jean Weisman, have welcomed me to their family and have been enthusiastic supporters of the adventures that Lyn and I have had as a consequence of my career.

Supportive friends, too many to list completely, were there at many stages in my career. Some include Smokey, Berry Lyons, David Meeker, Melanie and Dennis Hodge, Steve and Annie Frye, Berrien Moore, Bill Joslin, Peter Lamb, Gary and Donna Weisman, Mike Hussey, and Michael Morrison.

I have had several mentors throughout my career who have helped form my professional career: Parker Calkin, Dennis Hodge, Hal Borns, Richard Goldthwait, Colin Bull, Terry Hughes, George Denton, Claude Lorius, Bob Bates, and Bradford Washburn.

It would not have been possible to produce this book had it not been for my colleagues from GISP2 (the Greenland Ice Sheet Project Two). Several were particularly important to the success of the program and resulting science: Michael Morrison, Mark Wumkes, Jay Klinck, Mark Twickler, Kendrick Taylor, Debra Meese, Tony Gow, Richard Alley, Greg Zielinski, Jack Dibb, Michael Bender, Peter Grootes, Wallace Broecker, Jim White, Michael Ram, and many others. Colleagues from the European GRIP (Greenland Ice Core Project) always offered friendship and stimulating scientific discussions: Bernhard Stauffer, Michel Legrand, Robert Delmas, Henry Rufli, Claus Hammer, J. P. Steffensen, Henrick Clausen, and many others.

GISP2 would not have existed without support from the Office of Polar Programs of the National Science Foundation and in particular the support and advice of Julie Palais, Herman Zimmerman, and Peter Wilkniss. The drilling expertise of the Polar Ice Coring Office at the University of Alaska (Fairbanks), in particular Mark Wumkes and his drilling colleagues, cannot be understated: Without them, we would not have even had an ice core. We would not have gotten to Greenland without the New York Air National Guard (Scotia, New York).

Non-GISP2 activities referred to in this book were supported by the National Science Foundation, the Environmental Protection Agency, and the Electric Power Research Institute. Logistic support for these activities is greatly appreciated and was provided by the U.S. Navy VXE-6 Squadron, the New York Air National Guard, the Polar Ice Coring Offices at the University of Alaska and the University of Nebraska, Ice Core Drilling Services at the University of Wisconsin, Icefield Instruments in Whitehorse, Yukon Territory, Canada, as well as Sherpas and porters from Nepal, porters from India, yaks from Tibet, and ponies from northern China.

I was fortunate to have met my co-author Frank White, with whom it has been a great pleasure to share this book. His insight and experience were essential, but most importantly we have been able to keep up the process of writing this book and remain good friends for the several years it has taken to develop the book.

Finally, I hope that what we have written in this book will make a valuable contribution to the public's understanding of climate change and change in the chemistry of the atmosphere. Future decisions by the public related to these issues will have an immense impact on the quality of human life.

FRANK: I must begin with my wife, Donna. I started working on *The Ice Chronicles* around the time we met, and I have worked on the book since, throughout all the years of our marriage. Donna always supported this work, never begrudging the time I spent writing rather than being with her and our family, and she made very useful comments on the drafts. This book clearly could not have been written without her continuing love and support, for which I am very grateful.

I also want to thank all the other members of my family, especially my children and stepchildren. Ruth and Josh have seen me through five books now, and Liz, Nikki, and Jen are beginning to see what it is like to have a writer in the family! Thanks also to my son-in-law, Jonathan, and my granddaughters, Jessica and Madeline. All my family members have been enthusiastic about my writing and my belief that this book might make a major contribution to the debate about climate change. In the past, I would have thanked my parents, Frank and Mary Anne, but since they have passed on, I want to acknowledge the love and support of Bob and Jean Lane, my father-in-law and mother-in-law, during the writing of this book.

I also want to thank my friends Allen, Bruce, John, and Nick, who have been there for me through the writing of all my books, including this one. My colleagues at the office also deserve a note of thanks for their continuing interest in my life beyond the job.

Finally, I think it is highly appropriate to acknowledge my co-author. This is really Paul's book, but he has always insisted on shared authorship of it. He also has maintained a wonderful optimism and equanimity throughout the seven years it has taken us to bring this project to fruition. I could not ask for a better colleague.

PAUL AND FRANK: The figures for this book were produced by Sara Kreutz. Her expertise in graphic arts and willingness to help with this project are greatly appreciated. We also thank Kirk Maasch for developing the concept for one of the figures and Jane Fithian, Ann Zielinski, and Karl Kreutz for valuable contributions to the figures. Thanks also to Ike Williams, Hope Denekamp, and Dennis Campbell for their help, and to Lynn Margulis for her comments and excellent foreword.

Finally, special thanks to Richard Abel, a thoughtful and helpful editor, and to all those at University Press of New England who have helped bring the book to completion.

The Ice Chronicles

FIGURE I.1. "The Ice Core Time Machine." The 52-foot-diameter (15.8-meter-diameter) GISP2 drill dome and the 100-foot-high (30.5-meter-high) drill tower constructed by the Polar Ice Coring Office, University of Alaska staff, and GISP2 scientists during the summer of 1990. The map inset shows the location of the GISP2 site in central Greenland. The national flag of Greenland flies midway up the tower. Buildings (left of dome) were used to house equipment and as shelter for some of the GISP2 crew. Flags on either side of the drill dome provided orientation between the dome and the main camp in case of "whiteouts" (zero visibility conditions due to blowing snow). The main camp was located 0.3 miles downwind from the dome to minimize contamination potentially produced by camp activities such as generators.

Introduction

ON 1 JULY 1993, AT 2:48 PM LOCAL, THE U.S. GREENLAND
ICE SHEET PROJECT TWO (GISP2) LOCATED IN CENTRAL
GREENLAND . . . STRUCK ROCK. THIS COMPLETES THE
LONGEST ENVIRONMENTAL RECORD . . . EVER OBTAINED
FROM AN ICE CORE IN THE WORLD AND THE LONGEST
SUCH RECORD POSSIBLE FROM THE NORTHERN
HEMISPHERE.

—Message from Greenland Ice Sheet Project Two
posted Thursday, July 1, 1993, 4:56 PM EDT

Breakthrough

It was the middle of summer back home. Our friends and families were getting ready to celebrate the Fourth of July, when they would be standing around in shorts, grilling hot dogs and dishing up potato salad, maybe even watching a parade down Main Street. There would be none of that for us. We stood on top of the largest mass of ice in the northern hemisphere, the Greenland ice sheet, and shivered. We shivered partly because it was cold, but also because we were anticipating what was about to happen.

After all, this was Greenland, a place that is normally freezing, especially if you are 2 miles (3.2 kilometers) above sea level, on top of the glacier that covers most of the largest island on Earth. So being cold was nothing new for us. We'd been there for several years, scouting drill sites and then drilling down into the ice, hoping to reach the bedrock beneath it that hadn't been disturbed for hundreds of thousands of years.

What was new and different that day was a sound that we couldn't

FIGURE I.2. GISP2 team members extracting an ice core in the drill dome.

actually hear—the *kerchunk* of a 65-foot-long (20-meter-long) custom-designed drill hitting rock.

Monitoring instruments told us that, after five years of hard work, we had accomplished something unprecedented, and that we could now begin to unlock the secrets of more than 100,000 years of Earth's climate history. If you had been there, you couldn't have heard the drill hitting bedrock either, but you certainly would have heard the roaring of the team members' cheers in the frigid air.

What had brought these otherwise normal people—engineers, drillers, scientists, and construction workers—to a site hundreds of miles from civilization, working for months at a time in uncomfortable conditions, driving deeper and deeper into the Greenland glacier until they had penetrated all of the ice and hit the rock below?

We were all lured to Greenland by one of the great detective stories of science today—the quest to understand global climate change. We wanted to know how the Earth's climate works and how it affects human beings; these questions had led us to stand on top of a glacier in Greenland when most of our friends and families back home were getting ready for Independence Day.

However, as exciting as July 1, 1993, was for me and my colleagues who are in the business of researching global climate change, it was only one step in a long journey that had begun years before and is still under way. In the years that have passed since our success on the glacier, we have analyzed the results of the Greenland Ice Sheet Project 2 (GISP2) and undertaken further expeditions. We've also experienced a new urgency driving our work, as unsettled weather devastates various regions of the world, tropical diseases move northward, and island nations nervously watch sea level rise.

"Global warming" and "climate change" are no longer the concerns of academics, scientists, and explorers alone. These have become common topics of conversation among the people of the world, and pose issues of the utmost importance to the international community, which is struggling to write treaties and devise common approaches to the challenge that climate change is hurling at politicians and statesmen.

As just one example, some members of the scientific community issued an appeal to the business world in late 1999 to take immediate action to ameliorate the effects of global warming, foreseeing "rapid climate change," and "environmental turmoil." In a joint letter to London's *Independent* newspaper, James Baker, undersecretary of the U.S. National Oceanic and Atmospheric Administration (NOAA) and Peter Ewins, head of the British meteorological office, wrote, "Ignoring climate change will surely be the most costly of all possible choices, for us and our children."

FIGURE I.3. Inside the drill dome on the day (1 July 1993, 2:48 P.M. local time) that GISP2 hit bedrock and the bottom of the ice sheet at 10,018 feet (3053.44 meters).

Abandoning any pretense of uncertainty about the link between climate change and human actions, Ewins later told BBC Radio, "We're now coming clean and saying we believe the evidence is almost incontrovertible, that man has an effect and therefore we need to act accordingly" (McLean, *Boston Globe*, 1999). Strongly worded as this letter from two eminent authorities might be, it was only the latest in a series of warnings from those who study the climate to policymakers and the general public.

Recently, people have begun to believe that extreme weather is striking more frequently. Baker and Ewins cited the flood that killed thousands of people in Venezuela as an event that might be linked with global warming (McLean, ibid.). In 1998, thousands experienced the dislocations caused by an ice storm that crippled much of New England and eastern Canada. No one could remember anything like it, and it was natural to ask whether such an event is a harbinger of a future that includes unsettled, extreme weather resulting from larger climatic forces at work. During the same year, a long, hot drought parched Texas over the summer. Is that the new norm for a region that considers itself accustomed to heat, but had entire communities praying for rain under the pressure of 100-plus degree Fahrenheit (37-plus degree Celsius) days week after week? And as we approached the new millennium, massive storms struck Europe with little or no warning, with high winds and flooding wreaking havoc across the continent.

With these events in mind, one of the most important issues facing humankind surely must be understanding how and why the climate changes, with or without human influences playing a role. If that might have been an open question when we hit bedrock in Greenland in 1993, the matter seemed more or less settled as the year 2000 dawned. Yes, the world seemed to say, we must commit all available resources to understanding climate change as soon as possible. But how best to deploy our capabilities? How do we go about this momentous task and find answers based on solid science and thoughtful analysis?

Climate Change and Social Change

As is the case with so many public policy issues that have scientific inquiry at their base, a few definitions are in order. What is the difference between *weather* and *climate*, for example? And what does it mean when we talk about the climate changing? Is there "good" climate change and "bad" climate change, and how do we distinguish between the two?

First, the main distinction between weather and climate is the time period over which observations take place, and the geographical extent of those observations. Weather can be defined as the behavior of the atmospheric system in a specific region over a relatively short period of time, such as a day or a week. A description of weather would of course include temperature, precipitation, barometric pressure, storminess, and any extreme weather event, such as a tornado or hurricane. Typically, we also refer to the weather locally, that is, the "weather for today" in Boston, New York, or Paris.

Climate, by contrast, is the behavior of the weather over longer periods of time—seasons, years, decades, centuries, and millennia. Climate is also typically considered on a more extended regional basis, that is, across continents or subcontinents. We might, then, talk about the "climate in Europe during the past 6,000 years," and note that in spite of what we have seen in recent times, it has been uncommonly warm and dry, with less storminess than might have been expected when compared with earlier eras (perhaps 20,000 years ago). Scientists have now begun to construct models of global climate in order to simulate its behavior and investigate the consequences of extremes. But these models require validation through evidence (data) that demonstrates what the limits of climate can be, the factors that control climate, and how fast climate can change.

In the past century, the science of meteorology has advanced dramatically, especially with the assistance of weather satellites in orbit around the Earth. Today, it's by no means uncommon to predict the weather several days in advance, and the ability to issue early warnings about severe weather events such as tornadoes, hurricanes, and blizzards has saved countless lives. Until recently, however, the dream of climate prediction was just that—a vision that no one expected to see realized in the near future, if ever. Further, actual records of climate change, recovered from instruments that measure temperature, wind speed, precipitation, and atmospheric pressure, barely extend back into the late nineteenth century in the populated regions of the Northern Hemisphere, and cover even shorter periods over the rest of the globe. How, then, would it be possible to determine trends and patterns, and assess change?

GISP2 and other scientific expeditions are transforming how we view weather and climate. For example, while climate was once viewed as a constant backdrop to an ever-changing weather pattern, we now know that rapid climate change events (RCCEs) can shift regional or even global climate very quickly, even within the space of a few years. With the long records of more than 100,000 years available to us from projects such as GISP2, we can also begin to frame theories about how the climate changes naturally, providing a foundation for predicting climate, eventually even at a local level. This understanding of natural climate also allows us to assess human influence on climate, the newest addition to the equation.

Current concern about *climate change* implies that any global shift in climatic conditions would be a dramatic anomaly, something to be feared and avoided, if possible. However, the perception now takes a new twist because our studies show that climate change itself is natural, and that tremendous shifts in climate have occurred in the past. This does not lessen the importance of human influence on climate, but it does explain why assumptions of a single control on climate (for example, greenhouse gas increase alone) cannot provide accurate predictions.

Current thinking has been colored mainly by the relatively warm and beneficial climate that has prevailed for the past 8,000 years—during the period known as the Holocene (the last 10,000 years of Earth's history). Because this tiny stretch of time encompasses the period when all the great civilizations have arisen and fallen, it is hard to imagine how natural extremes in climate could be important. But new information has become available, particularly since 1993.

Also, when we think of climate change, we tend to focus almost exclusively on *global warming,* and how humans are affecting the Earth's climate patterns. But global warming, albeit an important one, is only one facet of the climate change equation. An equally intriguing question is how climate influences the development of human civilization. We must also answer simple questions before we will be able to distinguish what is normal from what is anomalous: what really causes climate to change, how much has it changed in previous eras, are extreme events common, and have past civilizations really been affected in dramatic ways by climate?

Although it is a given that weather does change every day, it has long been comforting to think of climate as a constant. Unfortunately, this perception of a static climate is deceptive, based as it is on relatively few records of the recent past, dating back to the mid-1800s. Seen against a broader backdrop, climate is not static, and climate change is clearly *not* an anomaly, but rather a normal feature of life on Earth.

By the beginning of the past decade, scientists had realized that a new view of climate was needed, as indicated by a report issued by the U.S. National Academy of Sciences:

> researchers have established the existence of a climate system on earth that is characterized by complex integration and feedback. The sun and all parts of the earth—the oceans, atmosphere, masses of land and ice, all life, the inner earth—are parts of this system . . . Climate change is actually a continuous process. The changes range from slow and gradual to surprisingly fast and dramatic. (National Academy Press, 1990)

Records such as those unearthed by the GISP2 confirm that climate always has changed and always will, because the Earth is a dynamic system that is part of a dynamic universe. The question that we should ask is not whether climate will change, but rather, "How do human activities alter the patterns of climate change? Can patterns of climate change, in the absence of people, be scientifically documented? How will changes in the climate (natural or human-induced) transform our daily lives?"

Increasingly, the answer seems to be that human activities indeed alter natural climate, and that climate change will in turn have a greater impact on our daily lives than we might like to imagine.

The idea of a continuously changing climate is not new. The late H. H. Lamb, the eminent British researcher and author of *Climate, History, and the Modern World*, was among the first to assert that the impact of climate on human history has wrongly been discounted because of the tendency to view it as a constant. Lamb argued that climate has indeed fluctuated over time, and its shifting patterns have been tremendously important to human society. Lamb argued that we cannot begin to understand history separate from climate. He said, "in the major breakdowns of societies and civilizations, climatic shifts may often be found to have played the role of trigger" (Lamb, 1982).

Lamb believed, even on the basis of what was known decades ago, that climate change might be one of the most profound influences shaping human history. As the climate shifts, civilizations adapted to specific weather patterns may rise, fall, or respond creatively and unpredictably to the challenge. And while recent millennia have enjoyed a relatively mild climate, fluctuations within that period have affected the ways of life of entire societies.

For example, just about one thousand years before the GISP2 team stood on top of the Greenland ice sheet, explorers from Norway and Iceland landed on Greenland's coast and established several colonies. They went on from there to North America, five hundred years before Columbus arrived in the New World.

The adventures of these Vikings took place during a time cimatologists call the Medieval Warm Period (MWP), an era of slightly higher than normal temperatures in Europe and elsewhere (meaning that these temperatures were high relative to today's temperatures and warmer than anything in the past two to three thousand years). The Vikings are remembered as great explorers, a bold seafaring people, and they were that.

However, it wasn't just a proclivity for adventure that made them colonizers—it was also something as simple as the warmer climate of the MWP. If the weather had been significantly colder at the time of their expeditions, there would have been no Vikings, as we know them, because they could not have sailed the ice-blocked oceans as they did the open seas of the MWP. When the period known as the "Little Ice Age" (LIA) succeeded the Medieval Warm Period, the explorations into these northern regions ended, and the colonies were apparently abandoned. In most cases the inhabitants just disappeared, and to this day we don't know exactly what happened to them. A climate record for the period of Viking colonization in Greenland would certainly help to unveil the mystery.

It is quite clear that we take for granted certain aspects of life on our planet today, so what might be thrown into question if we suddenly faced rapid climate changes unprecedented in recent history? We may feel comforted to think that our technological prowess would allow us to overcome challenges like those faced by the Vikings. Certainly, today's technology provides the tools to predict destructive weather, such as hurricanes, floods, blizzards, and tornadoes. However, as anyone who has endured extreme weather events can attest, our dependence on technology may actually make modern society far more vulnerable to climate change than our ancestors were.

Fears about the impact of the Y2K "millennium bug" were certainly overblown, but they reflected a valid concern that a global society dependent on a finely spun web of interconnected technologies may be highly susceptible to Mother Nature's whims.

We are already seeing the political impact of perceived climate change. The wealthier nations want the poorer nations to avoid taking a path of industrialization that will put more pollutants into the air.

The less developed nations, in turn, believe that the developed countries created the problem in the first place, and now need to take more responsibility for fixing it. These political splits have made it difficult for the global community to come to terms with global warming.

The Ice Chronicles

The study of climate is clearly valuable and perhaps essential to human survival. But there are many ways to study climate, and you might still ask why those of us who worked on GISP2 chose to pursue our goal of understanding climate change by studying glaciers.

The answer is that glaciers, such as the one in Greenland, are the storage systems for what we call "the Ice Chronicles," records of environmental change, most notably of climate history deposited in sites around the world that stretch over thousands of years.

We need these long records for perspective. Weather that may seem extraordinary to us because we have life spans of only seventy-plus years might well be part of a natural pattern when seen against a backdrop of several thousand years. When New Englanders say, for example, that the winters were much harsher a century ago, they may be right, but that doesn't necessarily mean that the higher temperatures and fewer blizzards today are the result of global warming—or does it? The New England climate was most likely calm 10,000 years ago, then stormy 5,000 years ago, then calm again.

What is happening today could be an unremarkable result of the natural climate system "doing its thing." Without the records, we just don't know. How could we?

As we have already noted, records drawn from direct observations, such as temperature and precipitation, have been rigorously collected for only about a century in the northern hemisphere and for significantly less time in the southern hemisphere. Historical journals from places such as China document change in climate, but these are relatively rare and extend back at most 5,000 years.

It is impossible to make definitive statements about human impact on climate without a natural baseline of comparison. Moreover, this baseline must stretch over thousands of years to create a reliable record of how climate has behaved in the past when human impact was minimal or nil. Long before human beings even existed on the planet, a natural climate system evolved. This natural climate system still exists and some of what we see today is undoubtedly a result of that system's behavior. We must view as much of the climate record as we can to understand that behavior, and to understand "paleoclimate"—all of climate history that has transpired from the origins of the Earth to the current moment.

Natural climate is "forced" or driven by a variety of factors, some working on long time scales, others that are short. For example, Earth's orbital cycles around the sun cause changes in the heat received by our planet over periods of 23,000, 40,000, and 100,000 years. These long cycles can affect climate in various ways that will be discussed in later chapters. Similarly, changes in the radiation emitted by the sun or in the atmospheric and oceanic circulation, such as the El Niño Southern Oscillation (ENSO) are examples of the shorter cycles (see fig. I.4).

Glaciers can tell us quite a bit about all of these cycles because of the way they capture the composition and behavior of the atmosphere. We have used glaciers for more than a century to reveal changes in temperature or precipitation (see fig. I.5), but now we have developed ways to extract more highly resolved, robust environmental information.

Records of ancient weather are preserved when snow falls through the atmosphere and is deposited in extremely cold regions. As the snow settles on the surface of the glacier, it collects gases, dissolved chemicals, and particles that, when analyzed, reveal the composition of the atmosphere at that moment in time.

In areas where the snow rarely melts, such as Greenland or Antarctica, these storehouses of climatic information are preserved for scientists to discover thousands of years later by drilling ice cores. These become "Ice Chronicles," stories of the ancient climate that we can read in modern times. High elevation sites in the low to middle latitudes are also useful because the deposited snow at these sites is preserved as well.

In addition to captured components of the atmosphere, we also find evidence of volcanoes erupting, forest fires burning, changes in the distribution of deserts, and many other tales of Earth's dynamic environmental history. The ice cores brought up from the interior of a glacier represent a kind of climate "time machine." Ice cores let us travel back in time and describe the paleoclimate with some precision—the ice core climate record provides a vivid, high-resolution view (even showing specific storms, seasons, and years) of the environment at any given moment in history.

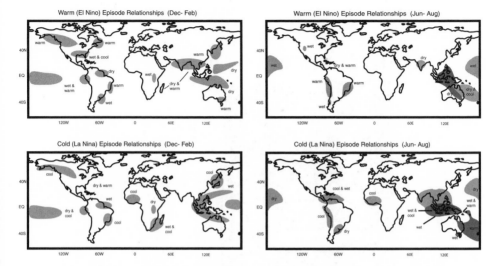

Warm (El Nino) Episode Relationships (Dec- Feb)

Warm (El Nino) Episode Relationships (Jun- Aug)

Cold (La Nina) Episode Relationships (Dec- Feb)

Cold (La Nina) Episode Relationships (Jun- Aug)

FIGURE I.4. Global map displaying typical anomalies in temperature and precipitation that coincide with the warm (El Niño) and cold (La Niña) episodes during December to February and June to August. For more details concerning ENSO, see chapter 6. *From the National Oceanic and Atmospheric Administration, Climate Prediction Center, National Centers for Environmental Predictions. Graphics slightly modified from www.ogp.noaa.gov/ENSO.*

FIGURE I.5. Change in position of glacier over time as shown in two pictures from the same glacier in Ladakh, Himalayas, with location map inset. The pictures are separated in time by 74 years. They demonstrate more than 0.6 miles (1 kilometer) of retreat of this glacier. The first (left side) was taken by the well-known mountain climber Fannie Bullock Workman during her 1906 expedition to Ladakh and the second (right side) by Paul Andrew Mayewski in 1980, when he led the first American glaciological expedition to the region. Arrow indicates same position on each photograph. Note in particular that snout (leading edge) of glacier is in contact with a rock cliff on the 1906 image but has receded from the cliff by 1980.

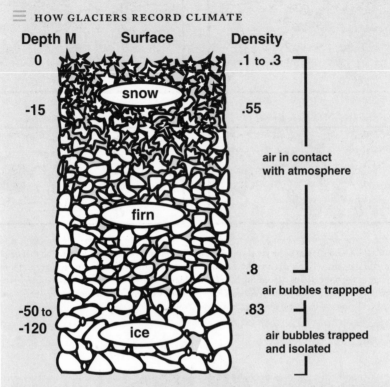

≡ HOW GLACIERS RECORD CLIMATE

FIGURE I.6. The transition in density (measured in grams/centimeter³ where water has a density of 1 gram/centimeter³) and in crystal shape from fresh surface snow to firn (old snow) to ice. The transition in density from fresh snow to ice is produced because of several factors: (1) Snow is added storm after storm and season after season, compressing underlying layers; (2) breakage and resettling of crystals plus melting leads to the gradual rounding of snow crystals; (3) compression of snow and ice by glacier flow is accomplished as snow and ice gets thick enough to deform under its own weight (typically several tens of meters in thickness).

The transition from firn to ice is accompanied by the closing off of air spaces to the surface. This process is important to the study of greenhouse gases in ice because the age of the gas is younger than the age of the layer at and above the snow/ice transition. Air still mixes with the atmosphere until this depth. In the case of central Antarctica, low temperatures and low yearly accumulation of snow slow down the density transition process, so gases can exchange with the atmosphere for hundreds to thousands of years.

Particles and substances dissolved in snow and ice do not exchange with the atmosphere under circumstances in which the glacier is in a very cold region (e.g., almost all of Antarctica and Greenland and high-elevation sites at middle to low latitudes). As a consequence, they are the same age as the layer of snow in which they were deposited originally.

If the snow that falls during one year does not melt or evaporate, then the next year's snow will fall on top of the previous year's layer. It is this process that creates glaciers. There are several types of glaciers. Those that fill valleys are appropriately called *valley glaciers*, while those in hollowed-out sides of mountains are called *cirque glaciers*. *Ice caps* are typically round in shape and encompass several square miles. The largest of the glaciers are known as *ice sheets*. In fact, only two ice sheets exist on the planet: the Greenland ice sheet and the Antarctic ice sheet (often divided into a West and an East Antarctic ice sheet, although they are in fact connected). We search for glaciers that are either in the polar latitudes or at high elevations, because both situations are characterized by minimal to no melting, thus preserving the record of years upon years of snow.

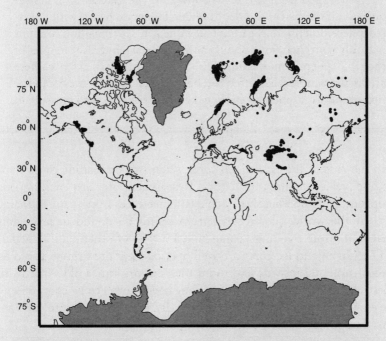

FIGURE I.7. Global distribution of glaciers. Gray regions are the two major ice sheets (Antarctica and Greenland) and black regions are smaller glaciers (e.g., ice caps, valley glaciers, mountain glaciers).

When snow falls on an ice sheet, it carries with it dust and chemicals from the atmosphere. In effect, the ice sheet has preserved samples of the atmosphere's chemistry from those times before direct records were taken, and has also stored them in sequential layers. As each layer of snow accumulates on top of previous layers, those below become compressed under the weight of the new snow being added at the surface.

When enough weight builds up, the snow turns into ice. This transition from snow to ice occurs at a characteristic depth for each site depending on the average annual snow accumulation and the average temperature. For example, the lower the temperature the deeper the transition, because the conditions that lead from the density of snow to that of ice are accelerated if temperatures are higher.

The depth where snow turns to ice is known as the *firn–ice transition.* A characteristic of the firn–ice transition is that small pockets of air are trapped from this level down into the glacier. As a consequence, air trapped at this depth can no longer exchange with the atmosphere. Therefore, ice from below the firn-ice transition contains not only concentrations of dust and chemicals, but also samples of the atmosphere from the time the bubbles formed. In essence, these are small "sample flasks" that can be analyzed to learn the composition of the atmosphere from times before we were able to measure it.

In the case of the Greenland cores, scientists are reading more than 100,000 years of climatic history. Researchers are also making our records more comprehensive by collecting ice cores from many different parts of the world, plus records from sediments found in lakes and oceans, and from tree rings, corals, and a variety of other natural media.

The Greenland ice cores are well over one hundred times as old as the instrumental records and more than twenty times older than the oldest written journals of climate. They have proven to be comprehensive enough to provide researchers with dramatic insights into the patterns that govern natural climate change.

Implications

What does it all mean? How do the Ice Chronicles help us to deal with the kinds of problems cited by those who see global warming transforming our way of life in the not-so-distant future?

First, we note that today's debate over the proper balance between environmental protection and human needs focuses on the immediate present and often polarizes the discussion by attributing all of climate change either to human actions or to natural climate patterns, when in fact the two are acting in tandem.

To some extent, this short-term either/or perspective is inevitable, but it is also unfortunate. Taking a longer view that includes a careful look at paleoclimate does not imply that human actions have no impact, nor does it mean that short-term policy actions are irrelevant. However, an expanded awareness ought to make a difference in how we approach policymaking.

At the broadest level, the Ice Chronicles have the power to transform our understanding of time, similar to the way that we have recently transformed our awareness of space. In the past thirty-five years, we have grown increasingly fascinated with our home planet, the Earth. What once was "the world" has been revealed to us as a small planet, a finite sphere floating in a vast, perhaps infinite, universe.

This new spatial consciousness emerged with the initial trips into Low Earth Orbit (LEO), and to the moon. After the Apollo lunar missions, humans began to understand that the Earth is an interconnected unity, where all things are related to one another, and where what happens on one part of the planet affects the whole system. We also saw that the Earth is a kind of oasis, a place hospitable to life in a cosmos that may not support living systems, as we know them, anywhere else. This is the experience that has come to be called "The Overview Effect" (White, 1987, 1998).

The planet Earth, which had seemed such a vast environment to our ancestors, now appears small, while the universe is overwhelmingly large. The Space Age has endowed us with a new spatial consciousness unlike any awareness available to humans in the past. Coupled with the technological changes made possible in part by communications satellites, also a product of the Space Age, this new global consciousness has changed how we make policy, expanding the process away from a national level and onto a global level. While some object to such globalization, it seems highly unlikely that we will ever go back to the past ways of doing things. It has also made us far more concerned with the environment, which in turn has led to many of the questions now being raised about climate change.

The systematic investigation of climate now under way has the power to imbue us with a transformed temporal consciousness similar

FIGURE I.8. Earth as seen from Apollo spacecraft.

to this change in spatial awareness. When we locate ourselves within natural cycles that endure for thousands of years, and witness the random events that punctuate these patterns, we open the door to describing what the recent past of climate history would have been without human involvement. We can then subtract this information from observed climatic activity and begin to uncover the human impact on climate. This new perspective, or "temporal view," should help us to decide not only what can be done to make things better, but also what is not worth doing. More important, it is another major step in our realization of how things work around us and where we fit into the natural system.

This book attempts to enlighten us about past history to help factor out the human impact on climate. In it, we report on the challenge, adventure, science, and policy implications that the Ice Chronicles offer to the world at a critical moment in human history, when climate has such practical implications for our daily lives.

We explore several basic insights drawn from studies of the Ice Chronicles that, taken together, radically shift our perception of climate change and how to live with it:

1. *Climate change should be expected as part of the natural world.* The Ice Chronicles establish a clear pattern of climate change before civilization had reached a level at which human activities might have any effect. Since both human and natural causes produce

changes in the climate, the question to be asked is whether human actions are countering or accelerating natural trends, and whether we are ready to cope with rapid climate change, regardless of what causes it.

2. *Human activity does have an impact on climate.* Definitive evidence now exists for unprecedented rates of increase and unprecedented levels (relative to the past million years) for byproducts of human activity, such as carbon dioxide and acid rain. The former produces warming of the lower atmosphere (troposphere) through the greenhouse effect. The latter, in the form of sulfuric acid derived from burning fossil fuels, can clearly cause regional cooling by shielding incoming solar radiation from the Earth's surface, as well as toxic effects to our ecosystem (Intergovernmental Panel on Climate Change, 1995).

3. *Quite dramatic natural climate change can occur in the space of a few years.* Through what are known as Rapid Climate Change Events (RCCEs), the Earth has already experienced massive reorganizations in the climate system. These are equivalent to plunging portions of the globe into year-long winters in the space of a decade or less and back into conditions similar to today's, also in a decade or less.

4. *The impact of current human activities cannot be analyzed without a clear understanding of natural climate, as seen over thousands to millions of years.* The long view is essential for many reasons that we have already discussed, and also because, if viewed over too short a time, a single event can appear to reverse rather than temporarily disturb a trend. The Pinatubo volcanic eruption, for example, cooled the entire planet and seemed to reverse a warming trend, but the effect dissipated rapidly, within one to two years.

5. *Certain natural cycles may play an important role in climate change.* Cycles as long as thousands of years and as short as a few years interact to help produce the climate that we experience. Interwoven with these natural cycles may be dramatic shifts caused by other, often chaotic events. These abrupt dislocations of the system are hard to predict.

6. *Current public policy is focused on reducing global warming, but we may also need to consider how to adapt to climate change with minimal social dislocation.* If all of our energies are devoted to policy disputes over preventing increased global warming, we may ignore the need

to cope with changes that are coming in the future, regardless of what we do today. We must understand the complexity of nonhuman versus human forcing of climate and the potential for dramatic surprises. Since climate change is natural, and we are part of the natural order, we should learn how to handle changes, while we try to minimize our contributions to climatic instability.

7. *We should also remain concerned about the impact on the quality of life produced by human-induced changes in the composition of the atmosphere through the byproducts of civilization.* Global warming is not the only issue that deserves our attention. Environmental changes of all kinds affect air and water quality and alter the landscape, while also affecting changing climate.

Let's now consider these issues in the light of the new insights provided to us by the availability of the Ice Chronicles.

1

Setting the Stage for Our Modern Understanding of Climate Change

In 1960, the United States entered a new decade with a new sense of energy and purpose. President John F. Kennedy began his administration with a call to explore a "new frontier" of opportunities, whether that meant securing civil rights for all citizens of the country, helping the poor in other nations create economic growth, or going into outer space. President Kennedy told us to ask not what our country could do for us, but what we could do for the country. And many of us responded.

As a young man, I found myself caught up in that spirit of adventure and discovery. I, too, wanted to do something that was not only exciting, but would also make a contribution to the world. Thus, my scientific career began in college (1964–1968) during a time of tremendous change and excitement in almost every area of human life, but especially in fields driven by science and technology. The seeds of one of these—the International Geophysical Year (IGY)—had been sown in the 1950s, but the impact would reverberate throughout the 1960s, exerting a major influence on me and on my field.

Scientists had conceived of the IGY as the first study of the Earth system in all its complexity, involving researchers from almost every nation on the planet. They planned to investigate the Earth from several different vantage points. This meant monitoring the behavior of the ocean, atmosphere, and processes on land, then combining this information to get more from the sum of the parts than would have been possible by single individual or even single country investigations.

The plan emerged from international scientific circles at the height of the Cold War between the United States and the Soviet Union, presenting a noble vision of international cooperation that would take place, ironically, at a time of intensified national competition.

The military and scientific dimensions of the Cold War merged in the IGY plan to launch the first artificial satellite of the planet Earth. Both the United States and the Soviet Union committed themselves to use their weapons of war (missiles) to launch a satellite that would be used for peaceful purposes.

On October 4, 1957, the Soviet Union sent *Sputnik,* the first artificial satellite, roaring into orbit on board one of their Intercontinental Ballistic Missiles (ICBMs). The "Space Age" had begun because of a program that was originally meant to focus attention on the Earth.

In response, the United States created the National Aeronautics and Space Administration (NASA) to coordinate all space-related activities and meet the challenge of the Soviet program. President John F. Kennedy gave NASA its primary mission: to land a man on the Moon and return him safely to Earth by the end of the decade. The United States achieved its goal in only seven years.

Ironically, just as the International Geophysical Year was intended to focus on the Earth, but instead shifted attention to outer space, so Apollo was oriented toward space, but turned our thoughts back to Earth. Some have even said that the Apollo missions to the moon triggered the ecology movement on the Earth. While this may be an overstatement, it's hard to deny that the two events were linked—Apollo 11 landed two astronauts on the moon in July 1969, and the first Earth Day was celebrated less than a year later, in April 1970.

In the words of shuttle astronaut Joe Allen, "With all the arguments pro and con for going to the moon, no one suggested that we should do it to look at the Earth. But that may in fact be the most important reason" (White, 1987, 1998).

The view of the whole Earth serves as a natural symbol for the environmental movement. It leaves us unable to ignore the reality that we are living on a finite "planet," not in a limitless "world." That planet is, in the words of another astronaut, a lifeboat in a hostile space, and all living things are riding in it together (White, 1987, 1998). This realization formed the essential foundation of an emerging environmental awareness. The renewed attention on the Earth that grew out of these early space flights also contributed to an intensified interest in both weather and climate.

The weather satellites launched after *Sputnik* are so common today that we take for granted their ability to deliver a picture of the Earth from space as a part of the nightly newscast. However, these satellites are the linchpin of accurate weather prediction, and their successors are laying the foundation for climate prediction. Earth systems science is one of the important new fields that has grown out of the revived focus on the Earth that resulted from the first human journeys into space. This new area of knowledge takes the whole Earth as its area of concern, with special attention to how human actions influence Earth changes, including alterations in climate patterns.

Interestingly, much of the study of Earth system science is now called *global change science*. Among the most important lessons learned from the IGY and earth systems science was the understanding of change in the system from local to regional to continental and global scales, hence global change science.

✳ **THE WRIGHT STUFF**

I note with much satisfaction that the talks on ice problems and the interest shown in them has had the effect of making Wright devote the whole of his time to them. That may mean a great deal, for he is a hard and conscientious worker.

Scott's Last Expedition: The Journals of Robert F. Scott, Beacon Press, 1957

I stood at the edge of the Wright Valley in East Antarctica and looked across the barren, boulder-strewn terrain toward the mountains on the other side. My thoughts drifted, far from the place I stood, a place I had sought to explore for years. My mind lingered on the Apollo program and the lunar landings that would soon be taking place. The year was 1968, and while others were going to the moon, I was making my first journey to one of the most remote places on Earth.

Wright Valley is named after Sir Charles S. Wright, a member of Sir Robert Falcon Scott's ill-fated expedition to the South Pole (1910–1912). Wright was a member of Scott's scientific staff, and he never actually saw the valley that bore his name until years later. Wright survived the expedition, which had ended with the death of Scott and several of his colleagues, and in the 1970s, he was brought to see the place named after him before he died. At the time of my own journey there, it had only been routinely visited for a decade.

Wright is called an "ice-free valley," because unlike most of the rest of Antarctica, it isn't always frozen over and covered with snow. These ice-free zones are rare; they account for less than 2 percent of

FIGURE 1.1. Photograph of Wright Valley taken by Paul Andrew Mayewski on his first expedition in 1968 with inset map of Antarctica showing the location of Wright Valley. Wright Valley is one of the so-called "ice-free" or "dry valleys" located along the coast of East Antarctica. The terms "ice-free" and "dry" are relative to the rest of Antarctica. The region is ice free because of a bedrock ridge that prevents ice from the interior from penetrating into the valley. Therefore, although it is located along the coast, the bedrock ridge funnels ice around this area—hence ice-free. Antarctica is the driest continent on Earth because moisture-bearing winds from the surrounding ocean are pushed away from the coast by strong winds that flow off the ice sheet and because the cold air cannot carry much moisture—hence dry valleys.

the surface of Antarctica. They are ice-free either because they are higher than the surrounding ice or because they are protected from inundation by mountain barriers. I suppose I thought of the moon while standing there because the terrain is so close to that of the lunar surface that NASA was using the ice-free valleys as a training ground for the Apollo missions.

I also felt a more direct connection with the astronauts. Exploring outer space is an extension of exploring the Earth. We've been exploring this planet for thousands of years, perhaps millions, and the region beyond its atmosphere for only a few decades. But it's the same process. If you are an explorer, you feel a kinship with all your kind, past and present, on Earth or the moon, Robert Falcon Scott, Ernest Shackleton, or Neil Armstrong.

FIGURE 1.2. Surface of the Labyrinth in Wright Valley, Southern Victoria Land, Antarctica, with Parker Calkin standing in front of our tent holding a UB (State University of New York at Buffalo) flag made in the field. The Labyrinth is a fascinating series of dead-end channels located close to the interior of Wright Valley. It was carved by fast-moving streams at the base of the ice sheet when ice was more advanced in this region. *Photo by Paul Andrew Mayewski (1968).*

I wondered, even then, why Antarctica fascinated me so much, why that was the first place I had wanted to explore so much. That's been a lifelong question, and I don't know if I can answer it to anyone's satisfaction, even today. Maybe we can't answer, rationally, what draws us to people, places, or vocations. I really think the most profound parts of our lives elude easy explanation. But I can say this: Antarctica's fascination for all explorers must ultimately be driven by the talisman of all exploration—the lure of the unknown. There are still parts of this great continent we have yet to explore, and we never know what an expedition will reveal. That makes each and every trip exciting. Back then, when I was making my first visit, it represented the realization of a dream I had nurtured since my sophomore year in college, at the State University of New York in Buffalo.

That day in 1968, as I walked down into the valley, crunching my way across the frozen ground, I asked myself, "Where will I be twenty years from now? Is this my only trip to this place?" The answer, of course, turned out to be a resounding "No!"

I've been back to Antarctica many times, visiting some of the most remote parts of the continent, leading over-snow traverses to unexplored regions, climbing mountains that had never been climbed by a human being before. Incidentally, there is another benefit of exploring. If you spend enough time in a place like Antarctica, they'll eventually name something for you, and I had a mountain named after me: *Mayewski Peak* is not far from Wright Valley.

The Ice Core Connection

Although the dramatic launching of Sputnik became the dominant story of the International Geophysical Year, original plans did not envision that to be the most exciting objective of the project. IGY also featured an ambitious commitment to exploring Antarctica. That's one of the reasons we were there around the time of the moon landings.

In the 1960s, while the space race overshadowed so much of the news around the world, scientists studying the climate were beginning to worry about the processes that drive the Earth system, just as they are today. However, they foresaw a completely different set of problems at that time. They focused, not on *global warming*, but on *global cooling*. They thought it might be very likely that the world was sailing right into a new "Ice Age," and they viewed that possibility with the same anxiety as we anticipate a worldwide increase in temperatures.

They envisioned glaciers advancing into populated areas, blizzards striking with ever-greater frequency, and heating fuel running out.

With hindsight, we understand why those researchers thought as they did. Records available at that time indicated that periods of climate like the one we experience today occur only 10 percent of the time based on records extending back close to one million years. Their observations are still essentially valid, but the question of scale is important. With some thirty years of perspective, we see that our predecessors' views offer a cautionary example for us as we worry about global warming. They were basically right, we believe today, but the timing for the next ice sheet invasion over North America is actually thousands of years in the future. In the meantime, faster changes in climate offer far more concern. This example also reemphasizes why extended records of past climate that are highly detailed (down to years and seasons) and that are representative of the whole planet are needed if we are to have any hope of understanding future climate.

The history of monitoring the Earth system as a whole is not very old. The IGY was the first major climate monitoring effort since the 1880s. Therefore, any results from IGY would be limited in what they might reveal about long-term trends. The only answer was to continue the monitoring begun during IGY so that, within a few decades, a record of sufficient length might be available to study change in the Earth system.

During the years immediately following IGY, monitoring was limited not only in terms of timeframe, but also in spatial extent, largely because of the cost of maintaining the monitoring systems. Unfortunately, if all monitoring takes place in, say, a region that's cooling off, it's only natural to think that your extrapolations to a larger scale might be biased. In fact, most of the monitoring during the 1960s was limited to the Northern Hemisphere and more specifically the North Atlantic, which was indeed cooling off at the time. This data should not have been globally projected, because we now know that this period was a short-term trend that only lasted from roughly 1940 to 1970.

Studies of past climate in other regions eventually yielded the information that turned this perception around. The first significant ice coring projects began in Greenland and Antarctica, as did interest in how the large ice sheets on these continents might affect, or be affected by, environmental change. Explorers undertook several large-scale traverses (surface crossings) of the Antarctic continent, and atmospheric monitoring in Antarctica and on a global scale began.

One of the pioneering efforts of the IGY era was led by Charles David Keeling at the Scripps Institute of Oceanography, the first in a series of continuous measurements of carbon dioxide (CO_2) (see fig. 1.3) and ozone levels that continue to this day, and that have become a matter of such great interest. At the same time, monitoring of ozone was begun at the British Antarctic station in Halley Bay.

How important was this work? It was vital, because the CO_2 monitoring alerted us to our current knowledge of greenhouse gas increases. It also led to the realization that the springtime ozone layer over Antarctica has been depleted in recent years. In 1995, the Nobel Prize in chemistry was awarded to Paul Crutzen (Max Planck Institute for Chemistry in Mainz, Germany), Mario Molina (Massachusetts Institute of Technology), and Sherwood Rowland (University of California,

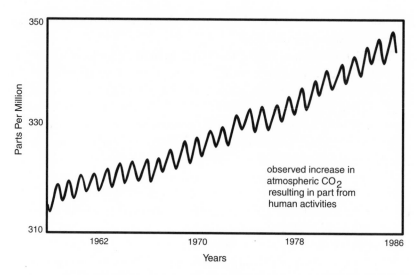

FIGURE I.3. Perhaps the most famous environmental monitoring record of the twentieth century, pioneered by Charles David Keeling of the Scripps Institute of Oceanography. When monitoring started in 1958, during the International Geophysical Year (IGY), the atmosphere contained 315 parts per million (ppm) by volume of CO2. By the 1970s, it had become apparent that a change had occurred. By the 1990s, the concentration had risen to over 350 ppm. *Data from Keeling, 1973; Rotty and Marland, 1986; U.S. Department of Energy, 1988.*

Irvine) for their research into the release of anthropogenic (created by humans) gases that cause the destruction of stratospheric ozone, called by some "the Achilles heel of the biosphere."

In spite of this auspicious beginning, however, much of the early monitoring turned out to be temporary, except for CO_2 and ozone measurements. Understanding why this happened is important in understanding how science is funded and implemented, and why some of the most critical activities don't continue.

Many of the original monitoring experiments were dropped when people realized that these programs were going to cost quite a lot of money, at a time when no one really expected to measure any significant changes. In particular, no one anticipated an upward trend in the atmospheric concentrations of CO_2—which is exactly what has happened, and which is at the heart of the global warming debate.

Moreover, many scientists—and some responsible for funding the work—harbored a bias against monitoring. They saw it as less important than other projects and not very "scientific." They thought it was "just data collection," which is shortsighted when you realize that you have to be patient through the early years of data collection to build up an information base robust enough to reveal something new.

Because these negative views dominated the debate, scientists who believed in monitoring had a real struggle on their hands to keep the experiments funded, and many of the projects were eventually terminated. For all its limitations, however, IGY's influence should not be underestimated. Certainly, the unintended consequences of Sputnik's launching have forever changed our views of ourselves, and the space program, especially satellite technology, has transformed our ability to understand weather and climate.

Moreover, new initiatives, such as the International Geosphere Biosphere Project (IGBP) reflect the exploratory spirit of IGY, while taking on many more areas of investigation. IGBP sponsors programs ranging from studies of paleoclimate to Earth systems in general, and from computer modeling to remote sensing satellites. Now, scientists are monitoring the greenhouse gases (e.g., CO_2, CH_4, N_2O) so important in global warming, largely because of discoveries that were made during IGY. The IGY opened up an incredible opportunity for worldwide information gathering about climate, and we would be far better off today if more of the experiments were still in place.

Initiation

I was very fortunate to become a scientist and make my first trip to Antarctica while still in my early twenties, at a time when all of these projects were under way. Many people helped me to achieve my goal; Parker Calkin was one. Now a professor emeritus at SUNY/Buffalo, Calkin is still an avid runner, and has always been physically fit. He was in his mid-thirties when I worked with him, and moved vigorously, almost gliding over the rugged landscape of Antarctica. I always envied Parker's physical stamina, and I emulated him, becoming a runner and triathlete myself.

Calkin's expertise is in the climatic implications of changes in land surfaces, especially glaciers, and he is officially known as a "glacial geologist." He taught me much of what I know about living in polar regions, and he also played a major role in sparking my interest in Antarctica. He had done his Ph.D. dissertation on the Wright Valley, and when I was a sophomore in college, he showed slides of his field site at a lecture. Something just happened inside me, and bang! Seeing those slides inspired me to want to go to that place as soon as possible.

Calkin was more or less responsible for inspiring me to go into the field and also to fall in love with Antarctica, but I suppose he still had to think about it when I asked him to take me there. Calkin is a careful scientist, but he was always willing to take a few calculated risks. He certainly made a great leap of faith when he listened to my constant pleadings and brought along an untested twenty-two-year-old to serve as his assistant. He gave me my first break, and my introduction into how to live, day in and day out, in the polar environment.

Of course, I didn't expect to go to Antarctica without some preparation. Two summers earlier, I'd worked with Dennis Hodge, a geophysicist, another of my mentors who also taught at Buffalo. He's only six years older than I am, and that had been his first year of teaching, as a newly minted Ph.D. Like Calkin, Hodge is very fit. (Are you getting the picture? You have to be in pretty good shape to do this work!) He's a tall, athletic man who looks more like a football player than a scientist. He and his wife Melanie were very good to me when I began my career. Hodge provided me with some of my first field experiences, teaching me the basic skills I would need as we tramped over the wild backcountry of Wyoming, making measurements that would help us to determine differences among buried rock formations so that three-dimensional maps could be reconstructed. I wrote my first published paper at his urging and learned a great deal about computer models (and punch cards, which we used then). He also taught me a great deal about the life of an academic—which also holds its own challenges to survival!

Then, the summer before my graduation from college, I worked as a field assistant with Calkin and Harold Borns from the University of Maine, mapping the extent of the glaciers that once covered New England. Borns is an experienced polar glacial geologist who also studies glacial deposits in New England and Scandinavia and founded the Institute for Quaternary Studies at the University of Maine. The legacy of Calkins, Borns, and the Institute stretched all the way back to the legendary polar explorer, Finn Ronnie. Working with them was my first initiation into a select group of people who have opened up these regions for the world.

After these apprenticeships, my dream had come true—I was in Antarctica as a fourth generation polar explorer, and pretty pleased with myself, to boot!

That's how it works in glaciology. In some fields of science, you can start to learn your craft in a fairly benign environment, and you might go through your entire career without being in any real danger. If glaciers are "your thing," though, you need to find mentors who can teach you not only the scientific side of the work, but also how to survive in the field. Slowly, and carefully, the existing corps of explorer/scientists initiates you into the mysteries of the practice. I'm doing the very same thing with my students now.

Each of us learns from others, and then we make our own contribution, which is a small part of the whole, then pass on what we know to the next generation. Sometimes, if we are very lucky, we have the opportunity to work on a project that really makes a difference, but even then, we are "standing on the shoulders of giants," building on the work of others. That, of course, is characteristic of science as a whole—each generation builds on the work of those who have gone before, and no single individual can say, "I did this!" without acknowledging the help of others.

I was also lucky because my original instincts proved to be correct, and I'm glad my mentors trusted me as much as they did. I flourished physically and psychologically in Antarctica and the other places that my career has taken me. I became a field leader myself within three years. In time, Antarctica became my second home, and it still feels very much like home for me today. I'm almost as comfortable there as I am in New England, where I've lived since 1975. From the beginning, it was just right for me. It's another one of those things that I can't easily explain, but I feel grateful for the opportunity to do something I love, and make a contribution to science as well.

✳ **POLAR REFLECTIONS**

Scientists like to focus on empirical, external reality, but I found that the quiet, vast spaces of Antarctica opened up opportunities for inner reflection that I couldn't ignore. It isn't just that the terrain hardly changes as you stand and look out over the flat white surface to the horizon, but also that it is so quiet. It's a different kind of sensory deprivation, I suppose. When there is very little external stimulus or change, we are thrown back on ourselves and our inner feelings.

Not everyone responds well to the situation. During some field seasons, we find that someone has a hard time with the isolation and the folding back on oneself that it causes. The best solution is to help them focus on their work.

There were instances when I'd come back to a place I'd visited in the past, and actually find an object I'd accidentally left there twenty years before (we try not to do that, but it happens). It would be untouched, resting quietly in the exact same spot where it had been left, freeze-dried in time! That seems unbelievable when you compare it with an American city, where nothing stays the same, or even in the same spot, for more than a few minutes.

Being in Antarctica also reminded me that all of the Earth isn't yet "humanized," and that as we continue to explore the universe, we will find similar places that help us to think in new ways about ourselves and our environment.

FIGURE 1.4. This hut was carried on board HMS *Terra Nova* and placed at Cape Evans in support of the British Antarctic Expedition 1910–1913. One of the primary goals of the expedition was to reach the South Pole. Captain Robert Falcon Scott, leader of the expedition, and four team members perished on their return trip from the South Pole. Inset map shows location of Cape Royds.

When the expedition left Cape Evans in the 1913, the hut was left with a large quantity of provisions, which proved useful to Sir Ernest Shackleton's Ross Sea party, who were marooned there in 1915 when their ship HMS *Aurora* was blown out to sea.

When I first visited Cape Evans in 1968, the hut was in very good shape but snow had blown inside parts of it (see inset photo). Historic huts like Cape Evans are now carefully monitored and maintained by the Antarctic Heritage Trust.

FIGURE 1.5. Moraines (ridges of debris) adjacent to a mountainous region called Robert's Massif mark the former position of the edge of the Shackleton Glacier (right center), a large glacier that drains ice from the interior of East Antarctica toward the coast. See Antarctic inset map for location. The moraines appear as ridges of boulders (left center) and larger ridges (center). The larger ridges, those closer to Shackleton Glacier, are younger (by several thousand years), and still contain ice, which causes the larger relief. The ridges of boulders are older moraines that date back at least several hundred thousand years. The boulder moraines look almost like the boulder piles built by the early settlers in New England. *Photo by Paul Andrew Mayewski (1970).*

When I first went to Antarctica, I was helping to reconstruct a history of the expansion and contraction of the ice sheet, changes that occur over tens of thousands of years as former margins of the ice sheet are captured by debris left at its edges. I wasn't really studying ice cores at the time. During my later explorations of Antarctica, I continued to marvel at the many unique aspects of the place, but I didn't realize for quite some time that the answer to many of the scientific questions that intrigued me were right under my feet. Ice cores, which later became the centerpiece of my scientific career, are the very stuff of the Antarctic landscape, and I was literally walking on top of my future!

Unfinished Business

During the late 1960s and early 1970s, environmental sensitivity grew worldwide. Intensified interest in the Earth as a system led to much greater curiosity about climate and where it fits into the environment. This growing concern about climate change coincided with the development of far more sophisticated computers, offering an opportunity to create global climate models (GCMs) for the first time. The dream of accurately modeling and even predicting climate change was born.

John Imbrie of Brown University pioneered the concept of holistic science with respect to Earth system science. He put together a unique coalition of institutions and conducted leading-edge work with the Climate Long-Range Investigation and Mapping Project (CLIMAP). Imbrie is a visionary who laid a solid foundation for those who followed him, especially in the area of interdisciplinary climate change studies. He is probably best known for his examination of long-term climate change and for demonstrating the important controls exerted on climate by Earth's position in space relative to our heat engine, the sun.

I was involved in CLIMAP as a postdoctoral student, working under a very well-known glacial geologist, George Denton from the University of Maine. Denton has dedicated himself to developing extremely precise reconstructions of the former extent of glaciers throughout the Arctic, Antarctic, South America, and New Zealand. Tall, intense, and eloquent, Denton has been working in Antarctica almost continuously since the late 1950s. His research has been instrumental in our understanding of climate change.

CLIMAP marked a milestone in climate research because prior to its inception, multidisciplinary research was far less common than it has become today. When Imbrie assembled his team, he was crossing disciplinary boundaries and persuading people to venture out of the safe harbors of their own fields of study. And it paid off. It was the combined efforts of oceanographers, paleontologists, soil scientists, geologists, and computer experts that made CLIMAP unique. CLIMAP amassed an amazing array of deep-sea sediment, glacial geologic, pollen, and soil records for its modeling and mapping effort and provided the first real "Earth systems" approach to the problem of climate change. Today, almost every project in this field can be seen as a descendant of CLIMAP.

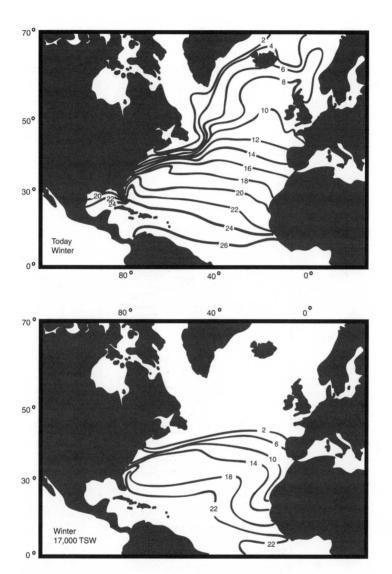

FIGURE 1.6. CLIMAP (Climate Long Range Investigation and Mapping) reconstruction of North Atlantic winter sea-surface temperatures during the last major expansion of ice sheets in the Northern Hemisphere, 17,000 to 23,000 years ago (lower plate) compared with today's situation (upper plate). Isotherms (lines of equal temperature) are used to reveal the dramatic changes. *Modified from McIntyre et al., 1981, Fig. 10-12.*

CLIMAP scientists established a new conceptual framework for our understanding of climate change. They demonstrated that the Earth's orbital cycles around the sun, which are produced by the gravitational attraction between the sun and its planets, exert a major influence on climate through changes in the radiation received by the Earth from the sun. They created reconstructions of climate as it existed 18,000 years

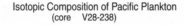

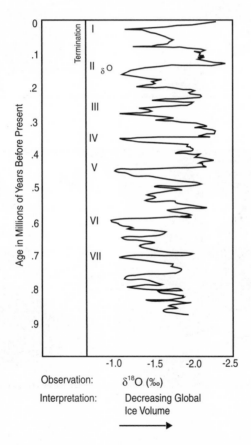

FIGURE 1.7. A record reflecting almost one million years worth of 100,000-year "cycles" of climate from a Pacific Ocean deep sea core (V28-238). This oxygen isotope record reveals changes in ice volume. More negative values indicate return of the lighter isotope of oxygen to the ocean as ice sheets melt (lower ice volume). Roman numerals represent the periods of rapid (several thousand years) melting of ice sheets (terminations) that end a glacial cycle and begin an interglacial. *Modified from Shackleton and Opdyke, 1973.*

ago, during the last glacial maximum (the greatest extent of glaciers during the last glacial period, about 15,000 to 100,000 years ago), and compared those conditions with today (see fig. 1.6).

Imbrie and his colleagues also calculated that interglacial periods (times between glacial periods, for example the last 10,000 years) tended to last about 10,000 years out of the approximately 100,000-year climate cycles observed from the analysis of sediment cores recovered from the ocean floor (see fig. 1.7). This is especially important, because we are now living in an interglacial period.

In fact, the current "interglacial" has lasted slightly longer than 10,000 years, prompting the belief that we should be entering a glacial period. The fact that the globe is not cooling (or rather, is cooling only regionally for limited periods, as in the 1940–1970 Northern Hemisphere case) means that the greenhouse effect and global warming may be causing the climate to run counter to the long-term trends.

In the past two decades, our studies have suggested that this is indeed what is taking place, and the debate has flip-flopped from a fear that all the glaciers would be advancing into populated areas to a concern that they are going to melt and raise sea level worldwide. This is one aspect of the climate change debate that has already generated fierce political struggles.

A Heated Discussion

Global warming emerged as the most controversial aspect of climate change in the 1980s. The issue received its biggest boost in popular consciousness when a series of extremely hot summers struck the American Midwest in the late 1980s, searing fertile farmland with months of drought.

In a defining moment in the climate change debate, Stephen Schneider, now a biology professor at Stanford, at that time a scientist with the National Center for Atmospheric Research (NCAR), testified before Congress and asserted that global warming had indeed begun. For many, his testimony shifted the debate from whether the planet is slowly warming up to a question of how much, how fast, and how detrimental the trend will be. Scientists still debate whether human activities are causing global warming, but others have stopped talking and are beginning to call for action.

A 1997 Boston *Globe* article laid bare the extreme fears of those nations most vulnerable to the effects of climate change:

> For the tiny Marshall Islands in the Pacific Ocean, global warming is a matter of survival: If the rising temperature raises sea level by 2 feet (.6 meters) over the next century, as many predict, 80 percent of the islands will be underwater. (Allen, 1997)

In other words, if the sea level rises, the Marshall Islands would effectively be destroyed as a viable society. What may happen to the Marshall Islands could also happen on a smaller scale in coastal areas of larger land masses, such as North America. Thus, while global warming isn't the only environmental issue facing us, its potential implications are dramatic and easy to understand.

The same article notes that other nations strongly believe that global warming is behind the extreme weather they have been enduring:

> In equally tiny Antigua and Barbuda in the Caribbean, with fewer than 100,000 residents in all, officials are convinced that global warming is behind a surge in hurricanes in recent years, including a 1994 storm that wiped out virtually the entire economy. (Allen, ibid.)

There are problems in some of the simplistic and all-encompassing assumptions people are making about global warming. It isn't the cause of *every* extreme weather event on the planet. Yet, the fears about its impact have spurred us to begin asking some of the right questions, such as "Where are we in the climate cycle? Are these changes natural or humanmade; are they transient or part of a long-term trend?" Suddenly, the much-maligned measurements that were begun during IGY started to look more promising, as researchers turned to records of carbon dioxide concentrations in the atmosphere and temperature data to find some of the answers to suddenly pressing questions.

As the 1980s drew to a close, the scientific community moved toward a consensus that these questions could not be answered without understanding natural climate history. That realization spurred us to consider how we could create a long record of natural climate—and that led us to begin our planning for the GISP2 project.

Public fears and scientific consensus coalesced into an unstoppable momentum that spurred the scientific community and eventually brought me to Greenland and that cold July day in 1993.

≡ GLOBAL WARMING, GLOBAL WARNING

Those of us living in temperate climates take for granted that there is a certain—we assume unchanging—group of infectious diseases that we have to fight. Residents of nontropical countries are familiar with illnesses such as influenza, which can be serious and even life-threatening. But an article in the Boston *Globe* points out that in the late twentieth century, the "flu" has been joined by new diseases as public health concerns (Allen, 1995). We have seen pneumonic plague in India coming on the heels of a heat wave and flood, the Hanta virus killing sixteen people in the American Southwest after a long drought followed by heavy snow, and cholera bacteria killing nearly 5,000 people in Latin America.

According to the article, these outbreaks of previously obscure diseases may be the early warning signs that global warming is beginning to affect people with life-and-death impact. The article is based on the work of a physician named Paul Epstein, and his colleagues at the Harvard School of Public Health's working group on new and resurgent diseases. Epstein notes that we are now seeing malaria, formerly confined to the tropics almost exclusively, in Houston, Texas, and he is quoted as saying that "if tropical weather is expanding, it means that tropical diseases will expand."

Why is this happening; what is the link between warmer weather and disease? The major reason is that the breeding grounds for rodents, insects, and bacteria expand with the warming temperatures. While we tend to associate cold weather with illness such as colds and the flu, freezing temperatures are also a deterrent to other kinds of diseases, such as malaria and cholera.

Paul Epstein has become a good friend over the past few years, and we've compared notes on our mutual interests. I respect his work, and have no doubt that his group has identified one of the most important consequences of climate change. Because of its direct impact on human beings, it is also one of the most important reasons for us to take action now, both to limit our impact on the environment and to prepare ourselves to cope with these new challenges.

2

The Making of an
Ice Core "Time Machine"

In 1988, the U.S. researchers involved in ice coring presented a proposal to the National Science Foundation to spend $25 million on a project called the Greenland Ice Sheet Project Two (GISP2) over a five-year period. About half the funds were targeted to logistical elements of the project (aircraft, ice core drills, field laboratories), and half to the institutions (there were twenty-five by the time of funding) that would participate. Everyone knew that it was a substantial amount of money for this kind of research, but it was minuscule compared with other government programs. Going to the Moon had cost $25 billion, while the GISP2 budget amounted to only $100,000 per participating institution per year to support students, staff, equipment, travel, analyses, and supplies.

By now, we had a lot at stake. On this proposal hung the hopes and dreams of a growing number of scientists committed to using ice cores as a fundamental tool of climate change research. The scientists and technicians who had signed on to this project ranged from faculty with permanent tenured positions to researchers whose sole livelihood depended on support from peer-reviewed grants. The graduate students were eager to learn and become involved in what they hoped would be important careers in science. The undergraduates, in most instances, were getting their first real experience in research.

The biggest challenge in doing science today is providing a setting with continuity for people who have dedicated their lives to solving important scientific questions. GISP2 provided funding for five years of support in which all of these individuals would be able to exchange ideas and have an opportunity to make a contribution. The difficulties inherent in creating and holding together teams of people with specialized capabilities over time remains one of the great obstacles to conducting scientific research in the United States and elsewhere. So we waited to hear from the National Science Foundation, and with so much hanging in the balance, it was hard to be patient.

The National Science Foundation (NSF) had begun serious planning for GISP2 in the late 1980s, but it wasn't an abstract exercise conducted in the NSF's Washington offices. Government agencies undertake projects of this kind only if they have the backing of key people in the field of study. Part of the planning process, then, involved convincing ice core research leaders to sign up in support of the project.

In this case, convincing people to come aboard was not a problem. The fathers of ice core research, people such as Willi Dansgaard of the University of Copenhagen, had previously spearheaded the campaign to conduct ice coring on the Greenland plateau as soon as possible. Dansgaard's opinions carried tremendous weight in the international scientific community because of his earlier work, which included helping to determine temperature ranges in the past by analyzing ice cores. Dansgaard and his colleagues also had a long history of recovering and interpreting ice cores from the Greenland ice sheet. For example, their early work figured prominently in our understanding of the climate conditions that led to the colonization of and later disappearance from coastal Greenland by the Vikings.

Chester Langway of SUNY/Buffalo also helped to lay the foundation for this project. His support made a difference with NSF because he had led the early deep ice core programs in Greenland and Antarctica, and he was known for his ability to get the job done.

Claude Lorius, an elegant and articulate French scientist, had conducted pioneering analysis of the results from an earlier Russian/French/U.S. "Vostok" ice core recovered from Antarctica. Vostok was important because it suggested that a relationship indeed existed between the presence of greenhouse gases in the atmosphere and rising temperatures, so his expertise would be highly relevant (see figure 2.1). The data collected by Lorius's colleagues has been widely distributed throughout the scientific and policy communities. Former Vice President Al Gore often displays the Vostok carbon dioxide–temperature relationship as an indication of the importance of greenhouse gases. We have also been able to determine from the Vostok record the extraordinary increase in carbon dioxide relative to the last few hundred thousand years.

Hans Oeschger, of the University of Bern, best known for his contri-

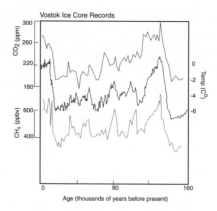

FIGURE 2.1. The graph shows data from the Vostok ice core in East Antarctica covering the last 160,000 years. Included are the greenhouse gases CO_2 (plotted as parts per million per volume [ppmv]) and CH_4 (plotted as parts per billion per volume [ppbv]). They are closely tied to Antarctic temperature variations over the last full glacial/interglacial climate cycle (Lorius et al., 1985; Barnola et al., 1987; Jouzel et al., 1987; Chappellaz et al., 1990). Temperature data are plotted as deviations from the present-day mean annual temperature at Vostok (−68.8 degrees Fahrenheit; −56 degrees Celsius). Comparisons like these are important for deducing the impact of changing greenhouse gases on climate. During glacial periods, atmospheric CO_2 and CH_4 levels were 30 percent and 50 percent lower than interglacial levels, respectively (Barnola et al., 1987; Chappellaz et al., 1990). Because CO_2 and CH_4 in the atmosphere absorb a portion of the radiation given off by the Earth's surface and re-emit it back to the Earth's surface (better known as the greenhouse effect), higher concentrations of these gases in the atmosphere contribute to a rise in global surface temperatures. The fact that atmospheric CO_2 and CH_4 levels were lower during the glacial period helps explain the lower temperatures and increased continental ice volume during these periods.

The Vostok ice core provides the longest continuous record of Antarctic climate available to date. These measurements have now been conducted through a depth of 3,350 meters (10,988 feet, or over 2 miles), approximately 420,000 years, covering more than three full glacial/interglacial cycles (Petit et al., 1997). Data from only the upper 2,100 meters (6,888 feet), corresponding to ~160,000 years, are included in this figure.

butions to understanding the carbon cycle and development of measurements of greenhouse gases from ice cores, rounded out this renowned team of senior scientists who were pushing the scientific community forward in search of the "ideal site" from which to recover the best possible ice core record in the northern hemisphere.

I knew that, if they were asked, these innovators in ice core research would support the American attempt to recover "the ideal ice core record." They had long argued that central Greenland would be the optimal site for major drilling, but they had been forced to work in northern and southern Greenland because those regions were cheaper and easier to access.

Their logic was compelling, but federal agencies must be cautious about spending large quantities of public money under any circumstances. However, the National Academy of Sciences (NAS) agreed with the idea that drilling on the central plateau made sense, which helped our cause.

Sleepless in Seattle

With the NAS recommendation finally in hand, the NSF told the ice core community to "get organized." Accordingly, the Ice Core Working Group (ICWG) held its first meeting in Seattle in 1986, and I attended. Little did I know what was about to happen!

About forty people were there, and we spent several days discussing how we were going to work together, holding the elections for chair of the group at the end. I was surprised, but greatly excited and enthusiastic, when they selected me. After all, it had been fewer than twenty years since I had traveled to Wright Valley for the first time as a first-year graduate student, and now I might have a chance to organize and oversee what promised to be a major event in science. I smiled, accepted congratulations, and shook hands all around. I tried not to show my concern, but thought about the phrase, "be careful what you wish for," because, after all, this was a great opportunity for science, and the greater the opportunity, the higher the stakes.

Anyway, there was no time for doubts now. With this new support from the National Science Foundation and the senior leaders in the scientific community, we began to work on GISP2. We also needed the help of senior U.S. scientists such as Wallace ("Wally") Broecker, of the Lamont Doherty Earth Observatory. He is a world-renowned geochemist who has been instrumental in understanding the chemical composition of the ocean and the significance of changes in this chemistry. He is also well known for his research linking changes in ocean circulation to changes in climate, a topic that was greatly advanced by GISP2 research. Wally provided constant scientific stimulation, advice, and support. We also wanted the backing of Charlie Bentley at the University of Wisconsin, who is a polar geophysicist who began his studies of the Antarctic during the days of the IGY. As a high-school student, I had pored over *National Geographic* pictures of researchers traversing the Antarctic ice sheet, and Bentley was of course prominently featured.

Years later, I found myself sitting on the same committees with him. The father of CLIMAP, John Imbrie, also came aboard. The project that would be known as GISP2 began to take shape and take its place in the history of science.

At the same time, our European colleagues were mounting their own ice core drilling program, called the Greenland Ice Core Project (GRIP). The leader of that project was Bernhard Stauffer, a professor at the University of Bern. Stauffer is known for his work on the extraction and interpretation of greenhouse gas records from ice cores. He is a careful and highly thoughtful scientist and has always been a great pleasure to deal with.

Debating in Durham

In 1987, the NSF's Division of Polar Programs sponsored a follow-up research workshop in Durham, New Hampshire, where a group of forty-five leading ice core researchers developed a strategy for recovering ice cores from central Greenland.

At that point, all we knew was that GISP2 would be an American expedition, and that GRIP would be led by a European contingent. The Durham conference also confirmed the value not only of drilling to the base of the Greenland ice sheet, but also the importance of confirmatory work in areas such as Antarctica and the Himalayas (EOS, 1988). That, of course, was also a dream for me, and I was excited to see it become part of the ice core community's basic approach.

We gathered on one of the hottest days in Durham, and yet we spent several days talking about snow and ice! We laughed about this juxtaposition during the breaks, perspiring in the sweltering New England summer, realizing that some of us might crave some of this weather soon!

Shortly after the Durham meeting, American and European executive committees of the two programs met to discuss precisely how we would cooperate. We made one of the most important decisions of the joint program then, which was to produce two parallel records instead of one record that we would jointly develop.

This decision was key because having two records would verify conclusively that any significant features, such as fast oscillations in temperature, were in fact a product of climate changes rather than simply being distortions in the sequencing of the ice preserved in the cores. These distortions could be produced by the natural flow of the massive Greenland ice sheet, for example, and might not indicate anything significant about real climate change. We spent a lot of time deciding how far apart we would drill the two cores to differentiate between environmental change and flow distortion, while staying close enough so that the two groups could help each other logistically.

This grand experiment, bringing up two ice cores separated by a distance of 30 kilometers (18.6 miles), would provide the scientific proof needed to demonstrate that phenomena such as rapid oscillations in climate really were caused by fundamental changes in the climate system.

Waiting for Washington

I left Seattle feeling excited and anxious. The Ice Core Working Group's goal was nothing less than mobilizing the U.S. ice core community—more than ninety institutions involved in various types of ice core research (twenty-five were eventually funded for GISP2) into a multi–institutional, multidisciplinary force capable of supporting the highly ambitious objectives of GISP2. These focused on recovering the longest possible record of climate change available from an ice core in the northern hemisphere.

We also intended to work at a site where that record would be so well preserved that it would allow continuous and accurately dated documentation of the environmental conditions affecting the region over the last interglacial/glacial cycle (approximately the last 100,000 years).

That was just the scientific challenge. The logistical problems facing us included constructing a camp, building a drill, designing field-based labs, and many other challenges—all of which were needed to recover and process the ice cores. Intellectually, I knew it could be done, but being responsible for overseeing the project was something else entirely. At the time, I was thirty-nine years old, and although I had had extensive experience in recovering shallow cores of 330 – 660 feet (100 to 200 meters) in Antarctica, the Arctic, and the high mountains of Asia, I hadn't done very much deep ice coring at all. In fact, U.S. involvement in most of those programs had ended years ago.

Once again, the expertise of the "old-timers" saved us. We had available to us senior scientists, such as Tony Gow, who had been involved in deep ice coring research from the beginning in Greenland

and Antarctica. Tony works for the Cold Regions Research and Engineering Laboratory (CRREL) in Hanover, New Hampshire, and he entered Antarctic research on one of those chance opportunities, just after finishing college in New Zealand. What luck that he did! He has been a researcher on more deep drilling projects than anyone else and always brings his great enthusiasm and vibrant energy to every expedition. I knew that I could depend on his counsel and that of others with his level of experience over the years ahead.

Just about everyone at the Seattle conference believed that we had a good plan and that approval for the Greenland expedition would be coming through eventually. However, we had also learned that you can never assume anything when you are seeking funds. Things happen that are unrelated to science and are totally out of your control, and you're left with reams of documents describing projects that will never happen.

In this case, we were right in believing that the funding would be forthcoming, but when we got the word, I needed every moment I had put into the planning and preparation, because the NSF gave us very little time once they made the decision. They had seen an opportunity for funding and needed to act quickly, so the pressure was on.

Detour to Antarctica

Regardless of what happened with GISP2, I had to go back to Antarctica to recover an ice core from close to my old stomping grounds, adjacent to Wright Valley on the Newall Glacier (see fig. 2.2), and to the South Pole (see fig. 2.3), to see if we could contribute to a newly identified environmental wake-up call—the so-called "ozone hole" (see fig. 2.4).

A satellite descendant of *Sputnik* had sent back disturbing data about a thinning of the ozone layer above Antarctica, and the first concrete signs of what came to be known as the ozone hole emerged from a British monitoring station that had been established back during the International Geophysical Year (IGY) in 1957.

Other satellites had confirmed apparent drops in springtime levels of ozone, recording a diminution in the number of ozone molecules in the atmosphere above Antarctica. The reason for the reduction appeared to be anthropogenic, that is, caused by human activities. Chlorofluorocarbons (CFCs), widely used in aerosol sprays, refrigerators, and air condi-

FIGURE 2.2. Tent camp on the Newall Glacier, adjacent to Wright Valley during 1985 ice coring program (see figure 1.1 for location). *Photo by Paul Andrew Mayewski (1985).*

tioners, seemed to be destroying ozone through a complex series of chemical reactions.

This discovery worried those of us in the scientific community because we understood the important dual role that ozone plays. When it is close to the Earth's surface, ozone functions as a "greenhouse gas," blocking the radiation that would otherwise escape into space, and raising the overall temperature. In other words, it contributes to "global warming." At the ground level, ozone is produced as a byproduct of human activities. It degrades air quality, and poses a health risk. Simply stated, this is "bad ozone."

But when ozone is high in the atmosphere, its natural role is beneficent, blocking dangerous ultraviolet rays, which can cause skin cancer in humans, from reaching the surface. This can be said to be "good ozone."

In the summer of 1988, the ozone was in all the wrong places and our concern spread to the general public. There was "too much of it in American cities, and not enough over Antarctica," in the words of Charles Oliver and Jake Page, authors of *Tales of the Earth* (Officer and Page, 1993).

The excess ozone hovering near the surface triggered another nation-

FIGURE 2.3. Excavating a 20-foot-deep (6-meter-deep) snowpit, 25 miles (40 kilometers) from the South Pole in 1990. It took four days to dig and collect samples from 600 levels in this snowpit at temperatures close to −58 degrees Fahrenheit (-50 degrees Celsius). We dug the pit far from Amundsen-Scott South Pole Station to avoid any potential contamination influence from the station. From this snowpit, we extracted samples that revealed several fascinating environmental stories. We discovered the first evidence of transport of Chernobyl nuclear disaster radioactivity to the Southern Hemisphere. We were able to verify that naturally occurring biological marine source chemicals can be used to track Antarctic sea ice extent and the behavior of the El Niño Southern Oscillation. We found evidence that contributed to the understanding of the Antarctic ozone hole. The researchers in the snowpit are dressed in "clean gear" to avoid contaminating the samples. *Photo by Paul Andrew Mayewski (1990).*

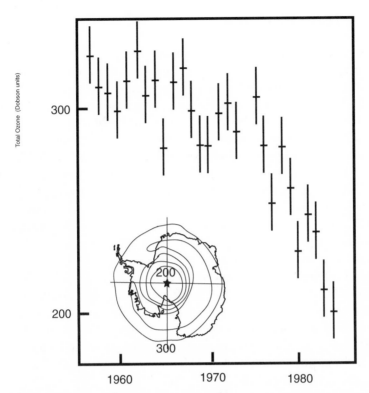

FIGURE 2.4. Antarctic springtime (October) total ozone data from Halley Bay Station, Antarctica for the period 1956 to 1984 based on ground-based measurements. Antarctic inset map shows contour lines for total ozone; note ozone hole centered over continent. Contours and October values are all in Dobson units (a measure of the total integrated ozone concentration in a vertical column from ground up).

The monitoring of ozone ranks as another of the most important monitoring experiments of the twentieth century. Like CO_2, total ozone concentrations have changed dramatically since the first atmospheric measurements conducted in 1957 during the International Geophysical Year (IGY). Note that levels of ozone (measured in Dobson units) over Antarctica in the springtime have decreased quite steadily since the 1970s. Ozone depletion occurs in the upper atmosphere over Antarctica, during the springtime when ozone-killing chemicals transported to Antarctica come in contact with sunlight triggering chemical reactions that destroy ozone. Some ozone-killing chemicals occur naturally in the environment, but the introduction of humanly produced CFC's has greatly enhanced the destruction process. This has lead to serious concern over the state of the ozone that shields incoming solar radiation in the springtime over Antarctica and adjacent regions and to a lesser but, nevertheless important degree, over portions of the high latitudes of the Northern Hemisphere. For more details concerning ozone, see figure 5.14. *Adapted from data in Farman et al., 1985.*

wide outcry, and helped to marshal support for passage of important amendments to the Clean Air Act of 1970. The thinning of ozone in the stratosphere, on the other hand, buttressed demands for the banning of CFCs, which were scheduled to be phased out in the near future by an international treaty signed in Montreal, the Montreal Protocol.

While the ozone problem over Antarctica would certainly have intrigued scientists under any circumstances, I doubt that it would have generated as much broad-based interest if it had been limited to that area. However, since the Antarctic findings were first made public, a similar thinning had been detected in Australia, New Zealand, and over North America and Europe as well. Now, everyone was clearly affected.

The alarm signals that sounded in the scientific community brought me back to Antarctica for more ice coring, but at a truly inconvenient juncture. I received the word on GISP2 in late 1988, as I was heading to Antarctica for the expeditions to the Newall Glacier and South Pole. I had a last-minute hunch, and thought that before leaving, I'd just put in one last call from California to the National Science Foundation. Herman Zimmerman, the NSF program manager, answered. I said, "Hi, Zimmie, I was just leaving, and . . ." He interrupted and said, "Paul, it's a go!" I thought, "Now what?"

The efforts of Zimmerman, Director of Polar Programs Peter Wilkniss, and other NSF program managers, played a major role in the success of GISP2. I can't say enough about their faith in our mission and the pragmatic political skills they brought to bear that helped make it a reality. "Zimmie" was the very first GISP2 program manager. He dove right into all the planning and dedicated himself to working with the scientists to produce the best possible record. He brought a deep devotion to the project and he carried it through as long as he was involved.

Wilkniss made sure that GISP2 never lost its priority status in the Office of Polar Programs, which required continued involvement at the scientific, financial, and logistical levels. His advice to me as leader of GISP2 was to "push on and do it right." He also counseled me always to consider how even a project as large as GISP2 could be synthesized into one or more truly important steps forward in science. This was sound advice for a project that was to include so many separate aspects that it might be easy "not to see the forest for the trees."

Partway through the GISP2 activities, Zimmie moved up in NSF and was replaced by Julie Palais as program manager for glaciology. Palais came from the scientific ranks where she had worked very successfully at unraveling the history of volcanic activity from ice cores. She understood GISP2 from the inside. It was also a great personal pleasure for me because she had been a student in several of the undergraduate courses I taught at the University of New Hampshire. I knew she would bring a great deal of talent and dedication to the job. We were fortunate to have had Zimmie and then to have Julie. A skipped beat at that early stage in GISP2 could have been disastrous.

Those with little experience in science often think of it as a clinical, objective, and emotionless process. The scientific method may well embody those values, but it is human beings who really carry out the science, and every program needs passionate advocates willing to fight for an idea to make it happen. Every program needs champions like Zimmie, Peter, and Julie.

Getting Ready for Greenland

Of course, I was happy that we were moving ahead, but the accelerated schedule meant that I suddenly had to figure out how to mount a field expedition to Greenland as early as the spring of 1989, and here I was, on my way to the other end of the world for several weeks.

Fortunately, it turns out that the best place to plan field expeditions is in the field! It gets pretty cold in New England, but nothing like the Arctic or Antarctic, and it's hard to visualize the real conditions you'll face when you're sitting in front of a roaring fire, knowing you can drive to the grocery store for food if you really have to. To plan a polar expedition, you must put yourself in a "polar mindset," thinking about how to handle every contingency with what is at hand.

So I headed away from Greenland and toward Antarctica. I was determined to use my time well, not only in carrying out the work at hand, but also looking ahead to GISP2.

Not long after I called Zimmerman on that warm California day, I found myself standing 25 miles from the South Pole. Being there was a goal that had cost many explorers their lives. For those who survived, being at the Pole was a supremely important "peak experience."

I wondered if the Norwegian explorer Roald Amundsen had shouted with joy as he trudged past this point on his way to a triumphant planting of his country's flag at the Pole on December 14, 1911. Or had Robert F. Scott of England and his weary men struggled through here on their desperate—and ultimately futile—attempt to reach safety after arriving at the Pole five weeks behind Amundsen?

We do know, from his diaries, what Scott felt when he arrived at a point near the Pole, found the remains of a camp, and realized that Amundsen had gotten there first:

> Tuesday, January 16, 1912: The Norwegians have forestalled us and are first at the Pole. It is a terrible disappointment, and I am very sorry for my loyal companions. Many thoughts and much discussion have we had. Tomorrow we must march on the Pole and then hasten home with all the speed we can compass. All the day dreams must go; it will be a wearisome return. (Huxley, 1957)

As someone who has been an expedition leader in Greenland, Antarctica, and several high mountains in the Himalayas, I have been especially moved in reviewing Scott's diaries. The Scott story is a tragic one. Having failed to be first at the Pole, he and the last of his men died in their tent, only a few miles from food and shelter. I knew how easy it is to lose yourself in the Antarctic wasteland, and I felt a moment of sadness for Scott. I wished he could have known that his death was not in vain.

✳ LOST!

Crawling on my hands and knees through the howling snowstorm, I didn't have time to be scared. But the thought did cross my mind that this might be my last expedition to Antarctica. I couldn't see a thing in the whiteout conditions of the blizzard; all I could do was follow the tracks of the sled and hope that John would realize I had fallen off and come back to look for me.

Antarctica has many ways to scare you, and getting lost in a blizzard is one that every explorer dreads. Scott and his party died in their tent only a few miles from their next food cache. Safety is our highest priority, but sometimes when you're trying to avoid one danger, you get trapped by another. This time, I was on a reconnaissance expedition during the 1974–1975 austral summer in an area full of crevasses, leading the first U.S. expedition to Northern Victoria Land. Crevasses are, without a doubt, the biggest problem in some regions of Antarctica. These are deep canyons covered by snow, which can give way without warning when you walk or drive over them. If you fall, you might plunge hundreds of feet, and if you aren't tethered to someone or something, you won't be climbing out.

I set up a system with my colleagues when I first started doing reconnaissance exploration so we could get across crevasses safely. We put the snowmobile in front, tied it to several sleds in back, with a climbing rope attaching the snowmobile driver to the sled driver. Working with a colleague, I invented a braking system that you could jam into the snow. If the snowmobile toppled into a crevasse, the person on the sled would activate the brake to help prevent a serious fall. You could then haul snowmobile and driver out of the crevasse. It was still scary, but it improved our chances of avoiding serious injury or worse.

The person on the sled has to stay alert, though, which is very hard to do when the wind is blowing at 60 miles (97 kilometers) per hour, the temperature is −6 to 14 degrees Fahrenheit (−21 to −10 degrees Celsius), and you're tired because you've already been out on the ice for 100 days.

Even with all my experience and knowledge of our sledging system, I fell off the sled in the middle of a blizzard. We were out in the storm because we had made one of those tradeoffs that Antarctica sometimes forces you to accept. We were trying to get to the shelter of a mountain range that was about a mile away. Normally, we would have stayed put because of the weather, but we knew that the winds were very strong in this region and the shelter of the mountain range could help to protect the tent. That made the extra travel sensible, even in the storm.

We weren't far from the mountains when we hit a bump, and I was thrown off. John Attig, who had been a graduate student at the University of Maine when I was a postdoc under George Denton, was driving the snowmobile. He didn't know that I hadn't had time to rope back onto the sled because the wind was gusting so hard that we couldn't hear one another. After about fifteen minutes, he realized I wasn't on the sled anymore and turned around.

John started back while I crawled toward him, and we literally ran into each other. I got back on the sled and held on!

We could see the mountain range off and on through bursts of blowing snow. It wasn't far now, but the snow was forming spin-drift, which means that wet snow gets inside the smallest openings, such as the sleeves of your jackets. It's unpleasant and dangerous because once it happens, you can never get your clothes dry.

We set up our tent and crawled in to wait out the storm. We thought our troubles were over, but they had just begun. We spent nearly three days huddled in our tent before the storm let up. It was cold, we couldn't start our tiny cook stove because the snow in our clothes would melt, and we had no way to dry them. We worried about frostbite. I'd heard that when you think you might have frostbite, you start speculating about which parts of your body you're willing to lose if you're allowed to get out of the situation safely. You kind of make a deal with the fates. I hadn't believed it, but I found myself doing it. I reluctantly decided a finger would be it for me.

On the day the blizzard let up, I said to John, "We've got to get back to camp or we're in a lot of trouble." We scrambled out of the tent and tried to start the snowmobile, but it wouldn't turn over—the carburetor was choked with snow. Base camp was 60 miles away, and we were hungry and cold. "See what you can do with it," I said to John, because he was a good mechanic. "I can't move my hands, Paul," he told me. My hands were freezing too, but I was at least able to move them a bit. "Talk me through it," I said, and after what seemed like several hours, we had gone through the whole process of changing out the carburetor, all the time hoping that the storm wouldn't blow over us again.

There's another secret to traveling in crevasse fields when you can't use the slow-but-sure method we'd developed with the brake and sled. It's only to be undertaken in extreme situations. You simply rev up the snowmobile and drive it as fast as you can. If the snow breaks under you, your momentum will usually carry you through to the other side. That's how we returned to camp on this occasion!

We were lucky. We got to camp safely without suffering frostbite or any other serious injuries. I was back at the University of New Hampshire for my first semester as an assistant professor just a few weeks later. Ironically, several years later, I was getting out of a car and closed the door on a finger. I actually lost a sliver of it. I'm a scientist, and I'm not particularly superstitious, but I did instantly make a connection between that incident and my "frostbite deal." It reminded me of how very fortunate I had been to get home in one piece. Perhaps I had paid a little bit of my debt to the fates.

Those early explorers had opened the way for us, just as we would for the expeditions of the future. That's the point of exploration. Some explorers succeed on their individual expeditions, some fail, but all give something vital to the overall enterprise of science. In a very real sense, no one fails.

However, the reasons for going to the South Pole in the days of Scott and Amundsen had been almost purely exploration and building national pride. Our team was there for much more practical reasons—we wanted to find out whether the ozone hole, which was directly above us at the Pole, was of recent origin. We also wanted to determine whether other important environmental phenomena could be detected using detailed records from the South Pole.

Antarctica is no longer simply the scene of races by explorers anxious to gain fame and prestige for their own countries. Rather, it has become an essential early warning system for the entire international community. As one example, the ozone hole is believed to occur naturally, but like greenhouse gas levels, its extent has been significantly altered by the impact of human activity. We were able to show that increases in nitrate levels (a proxy for a certain type of polar clouds) in the Antarctic snow indicated that the size and duration of the hole were probably greater than at any time over the last few thousand years. Nitrate is one of the primary constituents in polar stratospheric clouds, the depleted ozone appears to be related to the distribution of these clouds, and might be related to the size and timing of the ozone hole (Mayewski and Legrand, 1990).

One of the members of this Antarctic expedition was a close friend and colleague, Berry Lyons. Berry is a long-time co-worker of mine. He is a geochemist who has studied a variety of remote locations ranging from polar to desert regions of the world.

We spent fourteen highly productive years together at the University of New Hampshire, traveling all over the world, including trips to Antarctica, Greenland, and Asia. In fact, we ran together almost every day and then had lunch, discussing the day's annoyances, returning to work fit and ready for more. Berry is easy-going, and a delightful person to work with. Recently, he has led another major scientific effort in Antarctica, called the Long Term Ecological Research Program (LTER), which focuses on the ice-free valleys, and is dedicated to understanding how ecosystems work in polar environments. Projects such as the LTER demonstrate the emerging role that monitoring has in the Antarctic today. This monitoring now extends well beyond greenhouse gases and into detailed studies of ecosystems.

Jack Dibb, another colleague at the University of New Hampshire, was able to detect radioactive fallout from the Chernobyl nuclear accident from samples we collected on our trip to South Pole. Jack is an atmospheric chemist who, over the years, has led major efforts in atmospheric sampling worldwide. Most notably, he organized research in the center of the Greenland ice sheet at what is now the site of the former GISP2 camp for the purpose of monitoring the atmosphere on a continuous basis throughout a year, and for the first time ever through a winter.

This trip to Antarctica, which seemed to be something of a detour prior to GISP2, also provided me with the conceptual breakthrough that has dominated the rest of my career. It happened one day as I stood under the bright blue sky and looked out toward the stark white Antarctic horizon. Why the breakthrough came then I don't know, but an idea suddenly occurred to me—I realized that once we had the GISP2 data, we could pull together all the information from all the different sites, such as Antarctica and the mid-latitudes. That would give us worldwide coverage, as well as a long timeline, and we would be able to address many of the questions about climate change that were plaguing scientists, policymakers, and the general public. We would finally have the foundation on which to build comprehensive models of human impact on natural climate.

Why might GISP2 provide this foundation? Because we had all decided that it would be the most finely detailed and robust description of past climate and chemistry of the atmosphere ever compiled. We could envision this outcome because we had the backing of everyone required for success: the government, senior scientists, and an enthusiastic group of researchers, technicians, and students. What more could any project need—other than just a bit of luck!

The vision of accomplishing that goal drives me even today, and we are closer than ever to achieving it.

But I had much more to think about personally than the ozone hole, Chernobyl, or even the vision of a long climate record. I also needed to organize the Greenland expedition, so I spent my spare time in Antarctica getting ready for that. The trip turned out to be very useful in the end. I talked at length with Michael Morrison, who had just completed his master's degree at the University of New Hampshire. I told him our plans, and he said, "I can give you five years." I told him, "You're on; you just became the Associate Director!"

Trained as a mathematician and biogeochemist (a mixture of biology, geology, and chemistry), Michael brought a set of extraordinary skills to GISP2. His mathematical and technical capabilities, combined with his love for computers and an ability to interact effectively with people, made him invaluable. He has been everywhere and done everything. He had worked as a baker and a sheepherder before returning to school to get his masters, and he brought a "Renaissance person" approach to the program.

A former undergraduate student of mine, Mark Twickler, who had stayed on as a technician in my laboratory, was also accompanying me on the Antarctic expedition, and he just jumped in to take on a variety of tasks related to the planning of field equipment and laboratory space for GISP2. Mark had joined my research team after a history of doing construction work and had been with me on several field trips. In the field, he was tireless, working hour after hour, making sure that the lab and other facilities were ready for scientists to start working in the early morning. His efforts and those of Michael Morrison were truly Herculean.

We also needed a cook who could feed our crew and keep them happy through the long and arduous months of drilling. Since there isn't much to do when you're not working, and you need nutritious food to keep you going when you are on duty, one of the most important people in a program like GISP2 is the cook. You have to choose someone who is not only good at his or her craft but also has a sense of humor, because they will be the brunt of a lot of jokes, and will need all their ingenuity to come up with new dishes based on a very limited number of ingredients. I found just the person in Antarctica, Al Rosenbaum, and he signed up as well. I met him when he was the cook at South Pole. Since we were living in tents 25 miles away from the Pole, Al asked me what I would like to eat when we got back to camp. I said, "What would really be great would be to have a pizza delivered to us." Sure enough, when the vehicle came to our camp to drive us home, there was a pizza, piping hot in an insulated bag, provided by Al Rosenbaum. You have to hire a person like that, because that kind of a sense of humor is probably the most important trait in remote regions.

So I had an associate director (Michael Morrison), an assistant field planner (Mark Twickler), and a cook (Al Rosenbaum) by the end of my stint at the South Pole, and we had detected evidence of the residue of the Chernobyl disaster while doing our research on the ozone hole. We helped to show that the ozone hole is a modern phenomenon, and confirmed the first evidence of the Chernobyl nuclear accident traveling to the southern hemisphere (see fig. 2.5). This last piece of information, developed from our sampling near the South Pole, demonstrated just how sensitive the environment can be to individual human

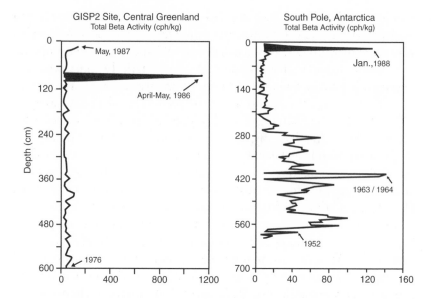

FIGURE 2.5. Total beta radioactivity (measured in counts per hour per kilogram [cph/kg]) from snowpits in central Greenland (GISP2 site) and 25 miles (40 kilometers) from the South Pole (Dibb et al., 1990). Both snowpits were hand-excavated to a depth of approximately 20 feet (6 meters) and then sampled for signs of radioactivity. The site near the South Pole contains snow dating back to 1952 and the central Greenland snowpit only back to 1976, demonstrating the greater amount of annual snowfall in central Greenland. Snowpits containing snow that goes back to the 1950s (such as the site near the South Pole) contain evidence of former atmospheric testing of nuclear bombs in the form of total beta radioactivity. These "bomb layers" provide a means for calibrating the exact age of the snow.

As of the mid-1970s, atmospheric testing of nuclear bombs was banned and radioactivity levels in the atmosphere and snow dropped to natural background levels. The 1986 nuclear accident at the Chernobyl reactor in the former Soviet Union released sufficient radioactivity to contaminate much of the high latitudes of the Northern Hemisphere (see highlighted black levels in central Greenland snowpit), but it was not assumed that this radioactivity could extend into the Southern Hemisphere. However, some radioactivity did rise high enough into the atmosphere to get into regions of the atmosphere where air can travel easily from high latitudes of the Northern Hemisphere to high latitudes of the Southern Hemisphere. The input timing for Chernobyl radioactive debris at South Pole is close to 19 to 20 months from the time of its injection into the atmosphere as indicated by the presence of high total beta radioactivity in January 1988 snow near the South Pole (note highlighted black section in snowpit). This dreadful accident provides a marker for the time of the accident close to its source and a fingerprint for tracing the time it took this air to reach the South Pole. The transport pathway is similar to that taken by ozone-destroying chemicals that are humanly produced in the Northern Hemisphere. Although most of their production is in the Northern Hemisphere, they turned out to have their first ozone-destroying consequences over Antarctica. For more details concerning the Antarctic ozone hole, see figures 2.4 and 5.14.

events. That finding further stimulated our interest in understanding how the atmosphere operates and what sorts of surprises might have affected humans in the past and could affect us in the future.

It was a very successful expedition, but the really hard work was about to begin!

≡ **DRILL SITE SELECTION**

All glaciers are not the same, and you can't just walk up to one, start drilling into it, and expect to get useful ice cores. Picking a good drill site means finding a place that meets several criteria to ensure that you will get well-preserved records.

Working conditions are usually harsh in polar regions, and the people who make a commitment to the project have to be tougher than the average person. Researchers trek to the very coldest regions of the Earth or to high-elevation sites in Asia and South America. I've been to the Himalayas and Tibetan plateau several times, and we often work within sight some of the highest mountains on Earth, most recently Everest.

Site selection usually begins with a regional survey, which in itself involves solving a host of logistical problems that would challenge the best military planner. In some areas, you can helicopter to within

FIGURE 2.6. A C-130 transport plane with skis at the GISP2 site. These aircraft can carry between 5,000 and 25,000 pounds of equipment to a remote site, depending on distance traveled and landing surface. The 109th New York Air Guard operate these aircraft for the U.S. National Science Foundation in Greenland. *Photo by Paul Andrew Mayewski (1991).*

FIGURE 2.7. The first U.S. traverse of Northern Victoria Land, Antarctica, led by Paul Andrew Mayewski in 1974 to 1975. Four expedition members traveled throughout Northern Victoria Land for more than one hundred days, carrying all of our food and fuel on four Nansen sleds, pulled by two snowmobiles. We produced maps of the area, and put out experiments to monitor ice movement that we reoccupied several years later. Note bamboo stakes on the last sled. These were used as trail markers in regions where we did not have mountains to use for bearing. Since the advent of lightweight global position satellite (GPS) equipment, this method of utilizing bamboo stakes for travel has been all but abandoned. *Photo by Paul Andrew Mayewski (1974).*

FIGURE 2.8. Tucker Snocats were used in Greenland during the 1987 reconnaissance expedition through central Greenland. The goal of the expedition was to gather information necessary for determining the best location for the GISP2 program. Expedition members used a dome tent as a kitchen. *Photo by Paul Andrew Mayewski (1987).*

a few miles of the proposed station site, or fly in on ski-equipped C-130 transport planes The United States is the only country with these kinds of planes, so the U.S. is an important player in many ice coring expeditions. Combined with these C-130s, we use snowmobiles or oversnow vehicles equipped to pull tens of thousands of pounds.

In Asian countries, the research team usually flies into an airport and then travels by bus, train, or truck until reaching the inevitable "end of the road." Then, we use horses, yaks, or other animals for the next phase, switching to human porters for the climb into elevations of 15,000 to 23,000 feet, or 4,571 to 7,010 meters. (As a comparison, consider that Mt. Everest, the world's highest mountain, is just over 29,000 feet, or 8,839 meters).

FIGURE 2.9. Yaks ascending the debris-covered Rongbuk Glacier with Mount Everest in the background. Yaks can carry loads of up to 100 pounds (45 kilograms) to elevations as high as 23,000 feet (about 7,000 meters). The Tibetan villagers who own the yaks travel several days from their villages during the spring and fall climbing season. *Photo by Paul Andrew Mayewski (1997).*

Once the team is at the site, we dig snow pits 2 to 6 meters deep (6.5 to 19.5 feet) to sample the ice cores; a 2-meter pit takes about a day to complete, while a 6-meter pit can take up to four days (see fig. 2.3).

We have to dig the pits by hand, because gas-powered engines produce local contamination, which is why it takes so long to finish them. This method allows scientists to analyze a wall of snow and observe the layering effects year by year. The retrieval team carefully

FIGURE 2.10. Mt. Everest from the north side with Rongbuk Monastery in the foreground, taken in 1997 on a joint Sino-American expedition to the region led by Paul Andrew Mayewski and Qin Dahe, a close friend of Mayewski's and member of the International Trans Antarctic expedition that crossed Antarctica in 1989 to 1990.

packs the samples for preservation and to keep them frozen, and then flies them back to the laboratory. Scientists working in the lab determine whether well-preserved signals clearly identify the different climate conditions in the years being studied.

The team also considers how spatially representative a site is, that is, how well it reveals climate history over a specific region or even the entire globe. They examine whether there has been much melting, and whether the annual snow layers can be seen, because this is important to understanding the timing of the record and preserving it.

For sites where we plan to conduct drilling, the team uses radar to show whether the relief under the glacier is flat or bumpy. Flat bedrock under the core site is by far the best for this kind of work because it doesn't distort the signals over the drill site. Steve Hodge of the United States Geological Survey (USGS) and his colleagues conducted airborne ice radar soundings over the region of the proposed GISP2 drill site to gauge the thickness of the ice and the conditions. During our ground reconnaissance (1987–1988) in search of a site for GISP2 drilling, the USGS crew flew countless sorties around the

FIGURE 2.11. Solar-powered drill operated during the 1985 expedition to South Greenland, funded by the Environmental Protection Agency. The expedition goal was to determine whether the "acid rain" produced by human activities traveled to remote portions of the Northern Hemisphere. The research confirmed this hypothesis. Once analyzed, the ice core revealed changes in acid rain levels that increased during periods of heavy industrialization and decreased once the Clean Air Act was passed. For details, see figure 5.17. Solar modules were used to prevent contamination during the drilling process that would be produced from a gasoline-powered generator. *Photo by Paul Andrew Mayewski (1985).*

clock to get enough detail to produce a map of the underlying bedrock and the ice thickness.

They tirelessly stared at their instruments and the ground, always searching for the right spot. We acted as a location beacon for them several times. Although they frequently flew over us, we never saw each other in person that season. They did an amazing job of determining the depth of the ice over the drill site, coming within a few meters of the actual drilled depth to bedrock: 10,018 feet (3,053 meters). Without the commitment and professionalism of Steve and his coworkers, GISP2 would not have been the success that it was.

When the best site is located after reconnaissance activities like these, the team returns with drills of various sizes and types, depending on the requirements of the project. For example, we used solar-powered drills in South Greenland for an Environmental Protection Agency project where the goal was to measure levels of pollution in the ice. A drill powered by fossil fuels would have contaminated the samples, which dictated the use of solar power. However, a solar-powered drill is limited in how deep it can go, and also by available energy that depends on the number of sunny days. There are also limits on how many solar modules can be transported into the field.

Small gasoline-powered drills can penetrate a few hundred feet or more into a glacier, while the really huge drills, such as the one that we used in the GISP2 project, had the potential to go well over 10,000 feet (3,048 meters). These monsters are powered by large 2 to 5 kilowatt generators.

Ironically, climate change is not only the subject of our research, it is also our greatest enemy, consuming many of the sites that could be used in these studies. Warming trends are already destroying glaciers in places such as Iceland, Asia, Africa, and New Guinea. I completed a survey of glacier fluctuations for Asia that helped verify the sad fact that by the early 1980s, many glaciers throughout that region were already on the verge of disappearing (Mayewski and Jeschke, 1979). It is in the very nature of the situation that these conditions will only become worse over time, which explains the urgency of our work.

Bringing Back the Ice Chronicles

GISP2 retrieved the longest continuous ice core record collected to date from the Northern Hemisphere, and the most detailed on Earth. The reconnaissance work for GISP2 had stretched from 1986 to 1988, and by late spring of 1989 a New York Air National Guard C-130 landed on top of the Greenland ice sheet. GISP2 was a reality. Additional flights brought scientific equipment, housing (huts and tents), plus food and fuel for a camp that would eventually house fifty or more researchers, ice core drillers, and camp staff for the next five spring-summer seasons.

Once we were set up, they flew in the drill, and I remember taking a minute just to stand and stare at it. I really appreciated the specialized skills of the people at the Polar Ice Coring Office at the University of Alaska who had built it. As tall as ten people standing on one another's shoulders, it was a monster that was going to chew its way through the glacier for just under two miles until it hit rock that hadn't been disturbed for perhaps 400,000 years.

Eventually, we'd need the esoteric thinking of climatologists, chemists, and mathematicians to read the message of the Ice Chronicles. But

FIGURE 2.12. The Greenland ice sheet and, at its center (star), the GISP2 ice core drill site (72.58° north latitude, 34.48° west longitude, elevation 3,207 meters or 10,519 feet above sea level) with idealized contours showing the position of the ice dome (Greenland's highest and thickest ice).

right now, we needed the brute force knowledge of mechanics and engineers, people like Chief Driller Mark Wumkes, an Alaskan mountain climber and jack of all trades, who pushed the drill ever downward season after season, and Jay Klinck, an experienced polar logistics manager. They would challenge the glacier and retrieve its storehouse of frozen "documents."

When we had started this process, we didn't have a deep drill, and we didn't have a proven process for cutting up the ice cores and preserving them. We'd built our camp over two field seasons, and had gotten all the people and equipment in place in just under one year. Going back to my Apollo analogies, it was a bit like landing on the moon because no one really knew how to accomplish that when President Kennedy said it would be done in seven years' time. However, given the support and

FIGURE 2.13. The GISP2 dome (for details, see figure I.1) and adjacent camp buildings that housed the galley and some of the team.

commitment of the government and the nation, NASA worked wonders. Now, with the backing of the NSF and the scientific community, we were attempting to pull off a few miracles ourselves.

As the drilling began, we hoped that all the talents of our colleagues in Alaska had been poured into the tool that had been flown halfway around the world to help us on our own challenging mission.

≡ **CAMPING OUT**

Dropping slowly out of the cloudless sky, a ski-equipped LC-130 aircraft operated by the 109th Air National Guard from Schenectady, New York, touched down on the surface of the Greenland ice sheet near Summit. Five people emerged, along with 9,000 pounds (about 4,000 kilograms) of navigation equipment and supplies. Within a day, this advance team had located the exact spot chosen for GISP2 drilling based on previous reconnaissance and had established the beginnings of the camp that would support scientists and drilling personnel. It was May 1989—GISP2 had begun.

During the summer months of the Northern Hemisphere, between forty and sixty people lived and worked at the GISP2 camp. The field season typically began in late April or early May and ended in late August or early September. From about May 9 to August 5 in central Greenland, the sun does not set, providing 24 hours of daylight. For most of the Greenland summer, temperatures at Summit range from 14 degrees to 32 degrees Fahrenheit (−10 degrees Celsius to 0 degrees Celsius), but may drop as low as −40 degrees Fahrenheit (−40 degrees Celsius) at the beginning or end of the season. Surprisingly, because of the cold temperatures, the air does not hold much moisture, and if there is no wind or cloud cover to block the sun, it is not unusual to find people working outside in tee shirts. New arrivals stepped out of the aircraft to a cold, flat, white world punctuated with the red and blue buildings of camp and the geodesic dome housing the drill. The camp was located at 10,449 feet (3,185 meters) above sea level, and newcomers found themselves gasping for air, even with slight exertion. When the wind is low and clouds are few, the "icescape" is brilliant. The surface of the snow sparkles with light reflected from snowflakes, and the horizon is utterly featureless in all directions except for the camp's buildings.

If there are clouds and the wind is blowing, snow fills the air and the line between sky and snow blurs. "Flag lines" along all major routes of travel around the camp prevent people from becoming lost during such "whiteout" episodes.

While annual accumulation of snow is less than one meter (3.28 feet), when a structure is erected above the snow surface, large drifts rapidly form during wind storms. Great effort must be expended to keep the buildings clear of snow. During the winter, we dismantled temporary berthing structures and stored them to minimize snow removal efforts when the field season resumed the following spring. Two permanent buildings were also designed to minimize drifting effects. The drill dome, because of its shape, induced a scouring effect, which greatly reduced snow removal requirements. The galley, known as the "Big House," was erected on stilts. Yet, in spite of all our precautions, the annual accumulation of snow had covered the Big House by 1995.

People slept either in "weatherports" or tents. Weatherports are red half-tube-like structures made from tubular aluminum arches and a canvas cover. With plywood floors and oil heaters, they can accommodate up to fifteen people. Some field team members who preferred more private, but colder, accommodations lived in unheated tents. Every year, a small "tent city" grew up with a thriving population. Everyone ate meals in the Big House. We all looked forward to "flight periods," when aircraft brought in mail and "freshies"—fresh fruit and vegetables. A diesel generator provided electricity for the camp, and also generated heat to melt snow for water. In fact, all the water used in the camp came from snow melted by the generator.

Life on the Ice

We spent much of our free time in the Big House, which also functioned as a conference room and dining area. Our work lives were spent outside for the most part, and many of the team members preferred to continue sleeping in tents. The drillers from the University of Alaska at Fairbanks rarely left the drill dome, unmistakable for its 110-foot (34-meter) tower poking out of the roof. The drillers got little relief from their duty in the dome, which was left unheated in order to keep the ice cores at less than −4 degrees Fahrenheit (−20 degrees Celsius), and to avoid disturbing its gas composition.

Mark Wumkes and his fellow drillers, such as Dave Giles, also an Alaskan, endured long hours and entire field seasons on the ice sheet's surface, setting up the drill and operating it, and ensuring that the best possible core could be recovered. Meanwhile, Camp Manager Jay Klinck made sure the constant demands of incoming aircraft and the details of camp life ran as smoothly and efficiently as possible.

≡ DRILLING DETAILS

Until GISP2, typical ice core drills produced cores no larger than 4 inches (101 millimeters) in diameter. Because of the number of researchers involved in GISP2, much more ice was required, and we developed a larger drill. The NSF contracted with the Polar Ice Coring Office (PICO, which provides logistic and drilling support for NSF-funded ice coring efforts), then at the University of Alaska/Fairbanks, to build a drill that would recover a core 5.2 inches (13.2 centimeters) in diameter, producing almost twice as much ice as the 4-inch drill.

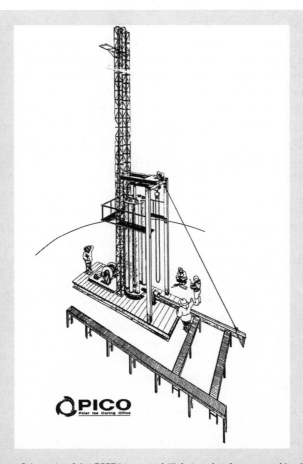

FIGURE 2.14. Schematic of the GISP2 ice core drill designed and constructed by the Polar Ice Coring Office (PICO) at the University of Alaska in Fairbanks. The drill tower (partially shown in this figure) stood 100 feet (15.8 meters) high and was housed in a 52-foot-diameter (30.5-meter-diameter) white geodesic dome. Several drill barrels were mounted on a rotating system to allow rapid removal of a barrel filled with one ice core and replacement of a new barrel. Barrels collected up to 18 feet (5.5 meters) of ice per run into the glacier. Run times (up and down) approached several hours once GISP2 neared the bed of the ice sheet. *Figure modified from engineering drawings provided by PICO, University of Alaska.*

With each run, the GISP2 drill typically produced 15 to 20 feet (4.6 to 6 meters) of ice core, 5.2 inches (13.2 centimeters) in diameter. The drill itself was approximately 60 feet (18 meters) long and was lowered down the bore hole on a 12,000-foot-long (3,657-meter-long) kevlar cable. The cable contained electrical conductors that brought power to the drill and returned information about the drill back to the operators at the surface. Because ice deforms easily under pressure, the hole formed by drilling would close on itself if there were nothing in the hole to supply back pressure. Many fluids have been used in the past to overcome this problem. A new drilling fluid, butyl acetate, first identified for this purpose by Ken Anderson, a chemist at the University of New Hampshire, was used in the GISP2 hole because it is one of the few fluids that has a low viscosity and low freezing point, and is relatively safe to work with.

The entire borehole was filled with butyl acetate to within 300 feet (91 meters) of the surface, supplying the necessary back pressure and keeping the drill hole open. The drill consisted of a cutter head, which actually cuts the ice, and a hollow tube, called the core sleeve, which accepts the ice core that is produced as the cutting proceeds. The section of the drill above the core sleeve is comprised of a pump that pulls drill fluid and chips from the cutter head around the outside of the cutter sleeve and into a screen section above the pump.

The chips are trapped in the screen section and the fluid returns to the hole. Above the screen section is the motor that turns the cutter head and the instrument package that controls the drill motor, and measures temperature, current to the motor, load on the drill, and speed of the motor. At the very top of the drill are three pieces of spring steel known as anti-torque blades. The anti-torque blades press against the side of the borehole to ensure that the cutters turn, cutting the ice rather than spinning in place and turning the drill instead. At the surface, a carousel holding a prepared core sleeve and screen section waits for the drill to arrive with a new section of core. The full core sleeve and screen section are loaded on the carousel, the new sections are loaded on the drill, and the drill goes back down the hole for another section of core. At a depth of, say, 7,382 feet (2,250 meters), the drill can produce 98 feet (30 meters) of core per day over two shifts of drilling. This means about 16 hours of teamwork every day.

Once it had been brought up into the drill dome, the new section of ice core was extruded from the core barrel, placed in trays, and cut into two-meter-long sections. To ensure accurate gas measurements, the

FIGURE 2.15. Mark Wumkes, chief ice core driller, Polar Ice Coring Office, University of Alaska, Fairbanks, inside the drill dome observing instruments built by Walt Hancock from University of Nebraska at Lincoln.

temperature of the ice was never allowed to rise above −4 degrees Fahrenheit (−20 degrees Celsius). During the summer, the temperature was often higher than this, so the ice was cut and immediately moved to the network of trenches below the snow surface. During part of this process, we actually stored the core in a freezer constructed beneath the surface of the ice.

The science trench stretched 15 feet (4.6 meters) below the surface, with side labs along it for scientists from different disciplines to do their specialized work on the cores. Here, the team cut the cores into samples where on-site analysis allowed us to do preliminary dating of the cores and gave us the first overview of the climate record.

The University of Alaska team controlled the drill from inside the dome through a computer linked to a microprocessor located inside the drill bit. Operators could "see" what the drill was doing by watching monitors that carried the signal from the microprocessor back to the dome.

We used interchangeable drill components, so that as one drill was bringing core samples to the surface, another was being prepared for

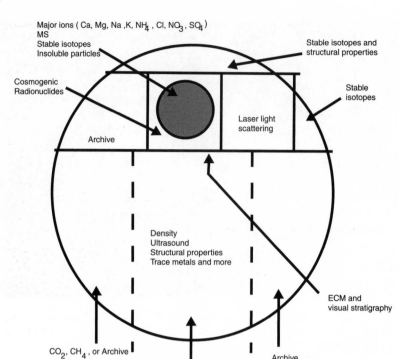

Major ions (Ca, Mg, Na ,K, NH$_4$, Cl, NO$_3$, SO$_4$)
MS
Stable isotopes
Insoluble particles

Stable isotopes and
structural properties

Cosmogenic
Radionuclides

Stable
isotopes

Laser light
scattering

Archive

Density
Ultrasound
Structural properties
Trace metals and more

ECM and
visual stratigraphy

CO$_2$, CH$_4$, or Archive

Archive

FIGURE 2.16. The ice core cutting plan. The 5.2-inch-diameter (13.2 centimeter-diameter) GISP2 ice core was cut into a variety of shapes to accommodate the unique requirements for each of the measurements. Some samples had to be extracted from close to the center of the ice core to prevent contamination (e.g., major ions, MS [methanesulfonate], insoluble particles), others required measurement along a surface (e.g., electrical conductivity (ECM), and others required a particular shape (e.g., structural properties).

lowering for further drilling. These hardened cutters returned up to 20 feet (6 meters) of material on each run. Our drill used up to 12,000 feet (3,657 meters) of 1-inch (2.54 cm.) cable and ran throughout the long hours of the Arctic summer.

On the "core processing line," each core was sliced along its length and parceled out to scientists representing various fields of study (see fig. 2.16), with half being archived for future uses, and stored at the National Ice Core Laboratory in Denver, Colorado. The lab is a joint NSF/USGS facility dedicated to the storage of these valuable archives. At the end of the core processing line, the remaining pieces of the ice cores were packaged in 6.6-foot (2-meter) sections in silver tubes nestled in plastic for shipment to the repository.

≡ THE CORE PROCESSING LINE (CPL)

Below the surface of the snow at the GISP2 camp, a 164-foot-long (50-meter-long), 10-by-13-foot-wide (3-by-4-meter-wide) trench connected to a large room where each core was stored before it was processed. This main processing trench was known as the core processing line (CPL). We kept processed core and samples in another large storage room before shipping them back to the United States. Most of the researchers actually left the field with prepared samples to be analyzed further when they returned to the United States.

The main CPL was relatively large but not enormous—100 feet (30 meters) long by 13 feet (4 meters) wide with six alcoves off to the

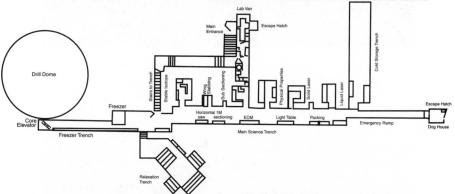

FIGURE 2.17. Schematic representation of the GISP2 ice core processing line (CPL). The research teams conducted more than 70 percent of all of the processing (cleaning and cutting of ice samples) in the CPL. Researchers, staff, and students spent up to 12 hours per day for several months each year in the CPL at temperatures close to −4 degrees Fahrenheit (−20 degrees Celsius). Although cold, the CPL became a place for lively discussion of initial results—a truly unique laboratory, classroom, and workshop.

Initially the CPL was located several feet beneath the drill dome, but as time passed and yearly snows accumulated, the CPL was buried farther and farther down from the surface. This necessitated that more steps be built each year to allow entry.

The CPL was divided into a series of side chambers, each dedicated to a different set of ice core measurements. A "relaxation trench" was excavated off to the side of the main CPL, allowing a place for ice recovered from depth (under high pressure) to adjust to lower pressure at the surface. Without a relaxation period (which turned out to be close to a year), the ice core would have been impossible to cut—it would have shattered.

Midway along the length of the CPL, samples were sent to surface labs that were warm enough to operate computers and analytical instruments.

An emergency exit and an access for pulling processed samples out and onto aircraft stood at the far end of the CPL.

FIGURE 2.18. This subsurface trench was excavated first by snow blower and later by electric chainsaw and by hand. The roof was made of plywood and supported by timbers. *Time-lapse photography of chain sawing and hand excavation by Paul Andrew Mayewski.*

FIGURE 2.19. Chris Kingma (graduate student) cutting ice core in the core processing line. *Photo by Paul Andrew Mayewski (1991).*

side. It was also very cold, with the temperature ranging from −22 degrees Fahrenheit (−30 degrees Celsius) in the early part of the field season to −4 degrees Fahrenheit (−20 degrees Celsius) near the end. An average of fifteen scientists worked ten or more hours per day in the CPL processing core. Although there was no wind in the CPL trench, working in the extreme temperatures required heavy clothing and occasional breaks to warm up.

The CPL played a critical role beyond simply processing the core. We performed many analyses directly on the core in the CPL, and discussion of these results among the researchers guided the course of further sampling and analysis. When one researcher discovered an interesting feature in the core, he or she would notify other researchers and they could then focus their efforts on the section of interest. It was an exciting time and a great way for young students to see real science in action.

These discussions helped to break up tasks that were otherwise tedious and cold, while also serving as an excellent training ground for students. The CPL was, in effect, a very cold classroom in which the first lessons from the Ice Chronicles were taught.

Core Issues

Three key criteria must be met to produce a valuable environmental record from ice cores. GISP2 did well on all of them.

1. *The record must be well dated over a long period of time.* Until recently, none of the records had been dated by counting annual layers of snow and ice for more than several hundred years. Our most experienced ice core researcher, Tony Gow, had been able to count the annual layers of snow and ice back almost 2,000 years on an ice core he had recovered from the South Pole, by seeing seasonal changes in the input of sea salt from the ocean. That feat had not been duplicated since, but we all knew that without a well-dated record we would not be able to apply the important statistical tests necessary to understand the record. With GISP2, we developed a new standard that employed multiple techniques for counting these layers and multiplied the length of the record some fifty times.

The first step was to identify the annual layers visually by shining

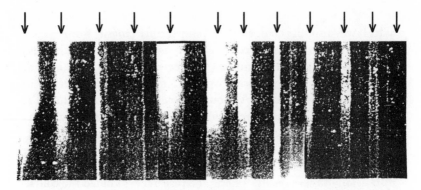

FIGURE 2.20. A 7.5-inch (19-centimeter) section of GISP2 ice core from 6,084 feet (1,855 meters) below the surface showing annual layer structure illuminated from below. This section contains twelve annual layers with summer layers (marked by arrows) sandwiched between darker winter layers. Lighter layers are produced by the increased scattering of light resulting from higher concentrations of dust that characterized this time of year during the glacial period. *Photo provided by Debra Meese (U.S. Army Cold Regions Research and Engineering Laboratory; Meese et al., 1997).*

light through the core. We could tell the difference in years either by changes in dust content (see fig. 2.20) or season-to-season alternations in the density of the ice (see fig. 2.21). Summer layers are less dense than winter layers because winds compact the winter layers. Moreover, during the winter-spring period, the strong winds blowing across Asia and North America carry dust into Greenland.

Several GISP2 scientists contributed to this aspect of the dating of the core. Debra Meese and Tony Gow of the Cold Regions Research and Engineering Laboratory in Hanover, New Hampshire, and Richard Alley of Pennsylvania State University tirelessly examined cut sections of ice. They worked 8 to 12 hours each day, staring at an ice core that, to many people, would appear to have nothing in it. But their trained eyes caught the subtle features that signaled change in seasons 1,000, 10,000, and 100,000 years ago. To be sure they had all seen the same signals, they counted and recounted.

Although they were highly trained scientists accustomed to more sophisticated science than counting, they realized that their skill was needed to make this record the best-dated ice core ever collected—essential if we were to answer important questions, once and for all, about the magnitude and speed of natural changes in climate. Only with such a record could we finally have a benchmark for comparing other paleoclimate records from less well-resolved or less easily dated sites. (Richard

Alley has written a very interesting book on his own GISP2 experiences, entitled *The Two-Mile Time Machine,* Princeton University Press, 1999).

Deb had been a graduate student of mine, having first done her master's degree work on the sea ice in Great Bay, New Hampshire, and then her Ph.D. on the sea ice off Alaska. It's a dangerous job, requiring constant vigilance, because cracks can appear quickly in sea ice. Also, even thin ice can look deceptively thick if the weather is poor, as it often is in the Arctic. Deb brought her dissertation research on the physical and chemical properties of ice to GISP2. She coordinated much of the detailed visual examination and the exacting task of counting the 110,000 annual layers identified in the GISP2 record. She displayed remarkable persistence, without which GISP2 would not have made its mark as substantively as it did. Richard Alley brought to the counting table the skills he had developed in Antarctica. Here he studied the stratigraphy of snow, examining its detailed structure in snowpits and

FIGURE 2.21. Snowpit near the GISP2 drill site. Alternating dark and light layers are related to the presence of relatively low- and relatively high-density layers in snow. The photograph was taken after two snowpits were excavated beside each other, leaving a several-inch-thick wall in between. One pit was covered (note plywood upper right of picture) and the other left open. Details of the stratigraphy within the covered pit are enhanced by light shining through from the other pit. *Photo by Paul Andrew Mayewski.*

ice cores. Of particular interest to him were the layers of snow that were very porous and of low density. These layers are called "hoar layers," and they are formed under conditions that promote transport of moisture from the snow pack to the air above or from deep in the snow to closer to the surface—as might occur during the summer when snow can be warmer than the air above.

This research eventually was combined with satellite work when Richard was joined by Chris Shuman, from NASA. Shuman and Alley worked out techniques to identify hoar layers and other features seen on the ground in satellite photography. The technique has enhanced the use of satellite photography as a tool for identifying the impact of climate events on snow surfaces and therefore the spatial distribution of climate effects.

Michael Ram, who came from the State University of New York at Buffalo, developed sophisticated instrumentation to scan the ice core at millimeter resolutions for changes in dust levels. A physicist turned ice core researcher, Ram developed new techniques for identifying dust content in ice using a laser. He eventually went on to use his data to demonstrate that the 11-year solar cycle of sunspot activity could be seen even in the 110,000-year-old GISP2 ice, demonstrating the constancy of this solar cycle.

Kendrick Taylor of the Desert Research Institute (part of the University of Nevada system) modified a technique developed by a Danish colleague, Claus Hammer from the University of Copenhagen, that gave us even more precision, at the millimeter level, in the search for annual layers. Taylor produced an instrument that ran an electric current through the ice core, allowing him to identify the electrical conductivity of the ice. This is important because high levels of electrical conductivity are related to the presence of strong acids, such as sulfuric or nitric acid. This gave us a good way to identify seasonal shifts in these acids, and it also helped us to first locate volcanic events, which are often characterized by the release of high levels of sulfuric acid.

Did the other scientists find Taylor's breakthrough useful? You can bet they did. Taylor worked almost nonstop, sharing his records as fast as he could get them off his computer in the CPL. We got the first glimpses of the jumps in climate from his records. These early views while still in the field, inspired us to think about the implications of these events. Every night, we gathered in the "Big House" to talk about the scientific papers we could write that would be based on these early views.

Seeing this data emerge almost immediately was an immense boost to us as we worked day after day on the drudgery of cutting up ice to be shipped to laboratories that would not see any results for months. Jack Dibb, my old friend from the University of New Hampshire, provided the reference levels for checking the fallout from nuclear bomb tests and the Chernobyl accident to give us absolute markers from the early 1960s to the present day. He was particularly expert in identifying these radioactive events, having studied them in several Greenland and Antarctic ice cores. Dibb still works at the old site of the GISP2 camp. He goes there year after year to collect samples of the chemistry of the atmosphere. With each year's data, he improves our chances of interpreting the GISP2 ice core record even more fully because he is helping us to understand the relationship between the chemistry in the air and that trapped in the snow and ice.

Greg Zielinski, then at the University of New Hampshire and now at the University of Maine, identified volcanic eruptions through sulfuric acid spikes. These could then be correlated with historical documents stretching back several hundred years. Since the early days of GISP2, Zielinski has gone on to develop the first 110,000-year-history of volcanic activity. He can not only determine that a volcanic event has occurred but has also been able to chemically analyze the tephra (volcanic ejecta) so accurately it can be pinpointed to its "home volcano."

Michael Bender, who was then at the University of Rhode Island and is now at Princeton, working with his former student Todd Sowers, now at Pennsylvania State University, measured the properties of the oxygen trapped in the air bubbles of the ice. These measurements were important because oxygen, as a gas, mixes globally, and biological activity provides us with a special signature that is embedded in the oxygen. The signal changes over time, so two ice cores of equal age should each have the same biologically introduced oxygen signal. Previously recovered Antarctic ice cores, like the famous Vostok ice core that had turned the tide in the greenhouse gas controversy, were not as precisely dated as the annual levels in the GISP2 ice, but were well calibrated on time scales of thousands and tens of thousands of years with other records (notably deep sea records). These provided another calibration for dating GISP2 ice cores.

2. *The record must be of a high resolution and continuous.* Although previous paleoclimate records had covered hundreds of thousands of years,

none had been continuously sampled at the resolution that was applied by the GISP2 team. We sampled the record at least every 2 years over the past 10,000 years, at least every 15 years for the period from 10,000 to 45,000 years ago, and at least every 50 years from 45,000 to 110,000 years ago. It was unprecedented to have a well-dated, continuously sampled record with such resolution. When we saw what we had retrieved, we knew that views of the climate record never before imagined would now be available.

3. *The record must include a sufficient number of properties.* The GISP2 record was to be sampled by some twenty-five different research institutions, with a total of almost fifty measurements to be developed from any specific level. We worked out a "cut-plan" to take into account all the different measurements.

To make the situation even more complex, each measurement required its own specific sampling protocol. People touching the ice with their bare hands, for example, would contaminate the chemical measurements. We could only make greenhouse gas measurements on ice that was maintained below a temperature of −4 degrees Fahrenheit (−20 degrees Celsius).

We held a series of meetings and laboratory tests that evolved a plan for collecting all of the proposed measurements. These measurements in turn provided an environmental reconstruction of the gaseous, soluble, and insoluble constituents in the atmosphere above Greenland throughout the period being studied.

Hitting Rock Bottom

I really don't know how to describe my feelings on the day when, at last, our drill struck the bedrock of Greenland, so far beneath where we stood in the drill dome. When you are participating in an "historic event," it can be hard to decide how best to observe it. So we just did what people typically do when they want to celebrate. We broke out champagne, we cheered, and we pounded each other on the back. I was happy, yes, but I think I was a bit subdued because I knew that the real celebrations could not, and should not, begin until we had unraveled the mysteries that the ice cores held encased within them—which meant that the real work would be continuing into the next millennium. I wondered, though, if we would find the answers quickly enough to anticipate what lies in store for humanity as climate change becomes a more pressing concern. And would we have any cause for celebration in what we found?

As it has turned out, even preliminary examination in the field of GISP2's early analyses revealed exciting insights. Continuing analysis of these "Ice Chronicles" is producing more information on an almost daily basis, years after hitting "rock bottom" in Greenland. Many of the insights gained from the Ice Chronicles of Greenland are finding their way into the mainstream press now, and have appeared in several hundred peer-reviewed papers. As a result, we are being compelled to re-think our understanding of climate change in fundamental ways.

3

The Discovery of Rapid Climate Change Events (RCCEs) and the Realization that Climate Has Multiple Controls

Imagine our early ancestors living within a few hundred miles of the vast ice sheets that covered much of the high latitudes of the Northern Hemisphere some 12,000 to 70,000 years ago (see fig. 3.1). They were accustomed to cold winters and harsh winds, but every 1,500 years or so the winds became stronger, pushed violently out of the north, and the winters lasted all year. These conditions might have persisted for decades, even centuries, forcing our ancestors to move southward. The change was so rapid that it occurred within a lifetime, which at that time might have been only twenty to thirty years. In fact, it happened so quickly that it might have been noticed by people in some areas in less than two years. Imagine also the tiny population of humans living about 50,000 years ago, perhaps fewer than 100,000 (Asimov and White, 1990).

If only a small percentage of the population experienced an invasion of terrible winds choked with dust, isn't it possible that stories would have been passed from generation to generation, urging people to fear the "terrible demons of the north"? After all, the winds, in the minds of those early peoples, came from the mouths of demons and dragons. The Ice Chronicles tell us that a scientific reality lurks behind these myths.

Imagine also that today, the winds suddenly increased and the winters lasted much longer, so that heating bills quadrupled, airline flights were frequently cancelled because of storms, and ice and snow brought most

North American Ice Cover 21,000 to 7,000 Years Ago

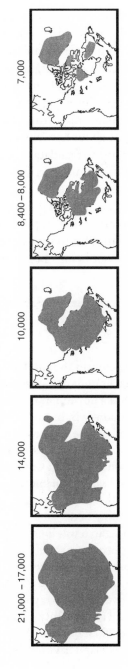

21,000 – 17,000   14,000   10,000   8,400–8,000   7,000

FIGURE 3.1. Changes in the distribution of ice sheets over North America and Greenland during the stages from maximum extent of the ice sheet (21,000 to 17,000 years ago) to the last major remnants of the ice sheet (7,000 years ago). *Modified from Mayewski et al., 1981.*

northern cities to a standstill ten or twenty times per winter. Now we have the scientific tools to determine the likelihood of such a scenario.

Since GISP2 hit rock bottom in Greenland in 1993, scientists around the world have been energetically analyzing more than 100,000 years of the Earth's climate history. So far, hundreds of papers have been published on the results of the expedition, and this work is likely to go on for years. GISP2, and subsequent supporting research, has already provided us with a raft of insights about how climate works. For the paleoclimatologist, the thrill of plotting real data on climate change over the span of many millennia is difficult to describe. It's a bit like looking at the Holy Grail of climate research.

For most people, however, GISP2 will probably be remembered, first and foremost, for its confirmation that Rapid Climate Change Events (RCCEs) are real, they are natural, and they have occurred many times in the past—even before human beings began to alter the Earth's climate system on a vast scale.

≡ END-TO-END SCIENCE

In 1970, on my second expedition to Antarctica as a graduate student, the National Science Foundation gave me the opportunity to show a U.S. Senator around a portion of the Transantarctic Mountains accessible from the main U.S. base, McMurdo Station. I had been selected for the honor because this was part of my research area. The senator and I spent a few hours together, and I felt comfortable enough with him to ask a question that was perhaps a bit odd, considering the circumstances. I said, "You know, I love working in Antarctica, but I wonder why the United States government supports the type of research we're doing here."

My point was that the research was incredibly exciting to us but it was not obvious to me why taxpayers should cover the costs. After all, it was unclear how the work would ever benefit the people who were paying for it. His response had a profound effect on the rest of my career. He said, "In twenty-five years, what you are doing will turn out to have practical value for all of humanity." I thought that was a very nice thing to say, but how could it be true? What possible value could research in the Antarctic wasteland have for the rest of the planet?

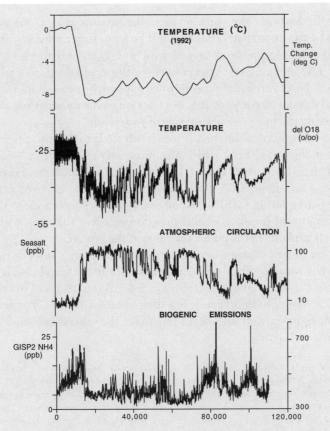

FIGURE 3.2. How the Greenland ice cores transformed our view of climate change. As recently as 1992, scientists viewed climate change as slowly evolving. For example, temperature change over a glacial/interglacial cycle (100,000 years) gradually cooled to a maximum (approximately −8°C global mean), then experienced a relatively fast (2,000 to 4,000-year) transition to modern conditions (upper plate). During our current interglacial (the last ~11,500 years, called the Holocene), temperature was assumed to have changed very little. Based on the foregoing, it was believed that natural climate variability operated very slowly and has been minimal during the Holocene. Thus it could be assumed that change in climate over the last century was not controlled by natural causes, but exclusively by human activities (notably an increase in greenhouse gases).

With the production of climate records from GISP2 and GRIP, it became clear that natural climate variability could operate very rapidly and occur frequently. Rapid climate change events with temperature shifts as much as −30°C affected central Greenland between 11,500 and 110,000 years ago (upper middle plate; data from Johnsen et al., 1992, and Grootes et al., 1993). These changes in temperature were accompanied by massive shifts in sea salt concentration, indicating dramatic increases in windiness over the oceans

While doubting the truth of what the senator said, I was still driven by the hope that I could *make* it be true. In recent years, this type of thinking has been tagged with a slogan: "end-to-end science." The meaning is, "Always look for the applied value in your research. Not every bit of it needs to be applied, but you should sort out and develop those parts that are relevant to the public while also conducting the more theoretical side of your work."

One more piece of the story brings it full circle to the information shown in fig. 3.2. When GISP2 was in its early stages, Peter Wilkniss, then director of polar programs, asked to meet with me. He offered me his support on what he knew would be the hard road ahead—completing GISP2. His advice was to try to represent the most important products of this program in a way that neatly and concisely explains to the public the value of the project.

Figure 3.2 and expansions on this figure in chapters 4 and 5 do just that. They revolutionized our view of natural climate variability. It is not slow. Natural climate variability, as it turns out, can lead to very dramatic change in significantly less than a human lifetime. The senator was right—we are now beginning to see the value of our work to everyone on Earth.

(lower middle plate; data from Mayewski et al., 1994, 1997, and Yang et al., 1997). Synchronous changes in dust levels (not on figure) signaled windiness over land (Mayewski et al., 1994, 1997, and Yang et al., 1997). The changes in atmospheric circulation over ocean and land have been far more subdued (note logarithmic scale on figure) during the last 11,500 years, but they still represent large enough changes to have influenced human activity (see chapter 4).

The Greenland ice cores also revealed dramatic shifts in terrestrial biomass seen through the measurement of ammonium in ice cores (lower plate, data from Mayewski et al., 1994, 1997). Increased levels of ammonium correlate to increased biomass. Extreme spikes in ammonium are related to biomass burning events in the modern record (Whitlow et al., 1994, and Taylor et al., 1996). These variations appear to follow the pattern of Earth's orbital cycles, notably the 23,000-year precession cycle (Meeker et al., 1997), demonstrating how closely plants are controlled by incoming solar radiation.

The Role of RCCEs

Studies conducted before GISP2 tracked a relatively warm period (an interglacial) some 120,000 years in the past that lasted about 10,000 years. This so-called "Sangamon Interglacial" (in North American terminology and "Eemian Interglacial" in European terminology), just slightly warmer than today, was followed by a relatively long period known as the Wisconsin Glacial (in North American terminology, or Wurm, using European terminology), very rapidly, so rapidly that it is called a climate termination. That then brought us to today's climate period, known as the "Holocene," covering approximately the last 10,000 years of Earth's history.

Until 1993, scientists confidently believed that the glacial climate operated slowly and that the Holocene was a quiet time for global climate, at least until the advent of global warming. However, that view is dramatically contradicted by the GISP2 information.

Earlier, expeditions to southern and northern Greenland had pointed to the kind of rich data that would eventually be gathered by drilling on the summit of the Greenland ice sheet. The "Camp Century" (northern Greenland) and "Dye 3" (southern Greenland) ice coring expeditions brought back tantalizing suggestions that what we call Rapid Climate Change Events, or RCCEs (pronounced "Rickies") might be a more common occurrence than anyone had previously thought. These results seemed to show that climate had moved from warm to cold very quickly, even in the space of a decade (Dansgaard et al., 1989). However, the idea of a static climate dies hard, and the data from these expeditions ran into some opposition.

Researchers questioned the records from Camp Century and Dye 3 because the rock bed beneath these sites was extremely bumpy. As a consequence of the uneven bed, the ice in these regions did not necessarily preserve layers one on top of the other. On the contrary, the ice twisted and even flipped layers over as it passed over the bedrock. Since bedrock effects can be translated well above the glacier bed, there was no unequivocal proof that the record recovered from these sites contained the original record as deposited. An ideal site to recover an ice core record would be a place were each layer is preserved as it falls and there is little if any disruption near the bed (see fig. 3.3).

Steve Hodge and colleagues at U.S. Geological Survey had demonstrated that the GISP2 bed was relatively flat and that we were literally on the summit of the Greenland ice sheet, a site from which all snow deposited descends down almost to the bed, vertically and without lateral flow (Hodge et al., 1990). The fast climate events we were attempting to confirm or disprove are often called Dansgaard/Oeschger events, after the Danish and Swiss scientific leaders of the expeditions that first found them. They are one type of fast change in climate seen in the

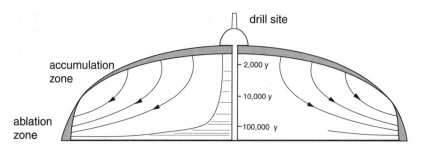

FIGURE 3.3. Cross-section (side-cut view) of Greenland ice sheet. An ice core drilled at the exact summit of an ice sheet captures each successive year of record trapped on top of the other. At any points off the summit, ice is carried away from the summit, in a downward and outward path. This path is produced by the driving force of snow accumulating on the interior of the ice sheet (accumulation zone) and a region of loss or ablation zone caused by snow and ice melting, sublimating (solid to gas), or breaking away as icebergs.

As ice is compressed year after year, it is forced so thin so that, in the case of the center of the Greenland ice sheet, the last 10,000 years of ice are in the upper half of the ice sheet and the remainder in the lower half. *Figure modified from original provided by Laboratoire de Glaciologie et Geophysique, Grenoble, France.*

GISP2 record, so we call these and other types Rapid Climate Change Events (RCCEs) in this book.

These early expeditions both ignited and fueled the debate over RCCEs. The questions they raised eventually led to the National Science Foundation commitment to GISP2 and the European commitment to GRIP (Greenland Ice Core Project). With two ice cores being retrieved from sites that were 30 kilometers (about 19 miles) apart, the results would prove conclusively whether ice flow had been responsible for the surprising interpretations of the earlier data. We reasoned that two sites separated by that distance would be in different ice flow regimes, so that if they both showed the same record, then the events must be true climate events rather than flow-developed events. As it turned out, GISP2 and GRIP agreed remarkably well.

The well-dated record that resulted from the GISP2 and GRIP ice cores changed forever and quite dramatically the view held earlier that natural climate variability operates on a slow time scale.

Producing this record was the true challenge. As described earlier, it took the construction of a fully functioning camp, the development of a totally new ice core drill, the collective energy of twenty-five universities, and countless hours of cutting ice in our natural laboratory, then analyzing the samples at home. A research team like mine dedicated a minimum of one hour to get a final measurement (combined sample processing, handling, analyses, data tabulation) exclusive of drilling time. Thousands of samples went into producing the data in figure 3.2.

☰ STABLE ISOTOPE MEASUREMENTS OF WATER IN ICE CORES

All atoms of the same element have the same number of protons in their nuclei and the same atomic number, but the number of neutrons may differ, yielding a different mass number. Versions of the same element with different mass numbers are called *isotopes*. Two important isotopes in ice core research are those of hydrogen and oxygen.

These isotopes are especially useful in providing us with information on the history of temperature. By far the most common isotope of oxygen is ^{16}O, with an atomic weight of 16. The next most common isotope of oxygen is ^{18}O, which has two additional neutrons in its nucleus and has an atomic weight of 18. Water containing ^{18}O, with the formula $H_2{}^{18}O$, behaves the same as water with ^{16}O chemically, but because it is heavier, it condenses more readily.

The ratio of ^{18}O to ^{16}O in water depends primarily on the temperature at the time the water condenses. By measuring the ^{18}O to ^{16}O ratio in ice core samples, we can calculate the temperature at the time the snow fell, because the heavier isotope is precipitated out as the temperature drops and/or it is precipitated out with distance of transport leaving more of the lighter isotope ^{16}O. Therefore, if a water molecule travels a great distance, enters a region of cold air (as, for example, it would when rising up onto an ice sheet from the ocean), then the ratio of ^{18}O to ^{16}O decreases.

This theory has been validated in two ways. First, researchers such as Willi Dansgaard (University of Copenhagen) and Claude Lorius (Centre Researche Glaciologie et Geophysique, France) have measured the distribution of surface values of snow in Greenland and Antarctica, respectively, and have compared these numbers to values for mean annual temperature at these sites (Dansgaard, 1964; Lorius and Merlivat, 1977). They concluded that essentially linear associations existed—the higher one gets onto an ice sheet or the farther inland, the colder it gets and the lower the ratio of ^{18}O to ^{16}O

because the former is precipitated out en route. Second, work by Curt Cuffey, formerly an undergraduate working on GISP2 and now on the faculty at Berkeley, compared measured temperatures in the GISP2 borehole to water isotope values in the core (Cuffey et al., 1995). He had to correct for the migration of the temperature record over time (heat diffuses through ice over time, so the record gets dampened). He found a strong association between the oxygen isotope–derived temperature and the measured temperature in the borehole.

The principal scientists conducting stable isotope measurements on the GISP2 samples were Piet Grootes, then of the University of Washington and now at Christian Albrechts University in Kiel, Germany; Minze Stuiver, from the University of Washington; and Jim White, from the University of Colorado.

☰ CHEMICAL MEASUREMENTS IN ICE CORES

The chemical composition of polar ice is made up almost exclusively of the following: chloride (Cl^-), nitrate (NO_3^-), sulfate (SO_4^{++}), calcium (Ca^{++}), magnesium (Mg^{++}), sodium (Na^+), potassium (K^+), ammonium (NH_4^+) and methanesulfonate (MS) (Legrand and Mayewski, 1997). In fact, these chemical species make up more than 95 percent of all of the chemistry dissolved in the atmosphere as a whole. Changes in the concentration of chemical species in ice are related to changes in source emissions and/or changes in the transport pathways of the chemical species to an ice core site. Snow and ice from Greenland may contain dust from Asia or North America, for example. It may also contain volcanic sulfate from, for example, Indonesia or Alaska, sodium and chloride transported in sea salt droplets from the mid-Atlantic, ammonium from terrestrial biota in, for example, arctic Canada, and biological source sulfate emitted by marine organisms.

Concentrations of these chemical species are measured in parts per billion (ppb) in ice core samples. Measurements at these levels are extremely easy to contaminate. As a consequence, researchers wear "clean suits" to keep their clothes and hands (we don't get to

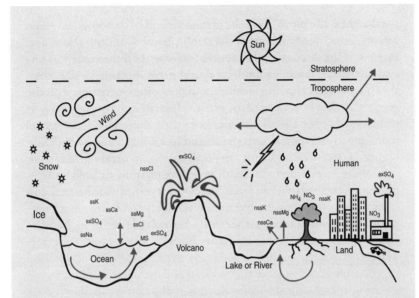

FIGURE 3.4. Sources for the chemical species transported to a glacier are numerous, but each tells an important story about either the strength of its source emission or the source and direction of transport of the air mass carrying the chemical species.

Chemical species that come from the ocean are transported either as seasalt (ss, typically as ssNa, ssCl, ssK, ssMg, and ssSO$_4$) or as biological emissions from the ocean surface (methanesulfonate, MS, or exSO$_4$). This ocean chemistry is incorporated into the atmosphere as wind blows over the ocean surface, incorporating sea spray, or as organisms emit gases from the ocean's surface.

Volcanoes emit gases that are often rich in exSO$_4$ (where "ex" stands for "in excess of seasalt SO$_4$") and non-seasalt chlorine (nssCl), allowing volcanic event histories to be developed from ice cores.

When winds blow over the land, they incorporate dusts in the form of nssK, nssCa, and nssMg. Stronger winds and/or arid regions allow more dusts to be incorporated into the atmosphere and their record to be trapped in ice cores.

Biological emissions from terrestrial plants, soils, and animals can be traced by viewing levels of NH$_4$, NO$_3$, and nssK in ice cores.

Byproducts of human activities can be seen in ice cores in a wide variety of chemicals, including exSO$_4$ from fossil fuel–burning and NO$_3$ from automobile emissions, plus trace metals such as lead, cadmium, and mercury (not listed on figure). *This figure was modified from one originally developed by Hans Oeschger, University of Bern, Switzerland.*

wash too often in remote regions) from touching samples. (Note the clean suits in figure 2.3.) The very low levels seen in central Greenland ice compared with measurement of these chemical species closer to their sources in ppm (parts per million) or ppt (parts per thousand) and higher is a result of the distance traveled (the farther the distance, the more likely the chemicals will be deposited en route by rain, snow, or dry fallout of the atmosphere). Concentrations also increase if the strength at the source increases or if the transport energy increases. As an example, if desert areas increase in size, then the source of calcium, magnesium, and potassium increases, with the result that levels measured in central Greenland also increase. If wind speeds over the desert increase, then these chemicals travel more quickly to central Greenland and less is lost en route, yielding higher concentrations on deposition in central Greenland.

During glacial periods, there is less vegetation to hold soils in place and stronger winds yield increased levels of continental and marine source chemicals. Even marine biological source methane-sulfonate increases because the oceans are turned over by higher surface winds, bringing more nutrients from the depths to the surface. Terrestrial source ammonium usually decreases, however, because vegetation on land is generally reduced under colder conditions.

Using the chemical species discussed above, we can also reconstruct atmospheric circulation patterns, because the chemistry acts like a fingerprint for the air mass (see the discussion of this phenomenon in chapter 5).

The results from Greenland demonstrated not only that RCCEs really do occur, but also that they occur far more frequently than anyone had thought. In addition, as we discuss in more detail in chapter 4, RCCEs even occur during relatively mild periods such as the Holocene interglacial. We can therefore no longer write off the Holocene as an era with little or nothing new to tell us. The Holocene has in fact been hit, hard and often, by climatic transformations.

Overall, we have learned a great deal about climate variability on the scales of decades to centuries and even as finely honed as seasons to years, because of the unprecedented changes in climate recorded in the two ice core records from GISP2 and GRIP. Throughout the remainder of this chapter, we will examine RCCEs that operate on the millennial scale.

RCCEs, which are really massive reorganizations of the climate system, were most extreme during glacial times, 10,000 to 70,000 years ago, when Northern Hemisphere ice sheets provided a positive feedback (reinforcement) to the climate system, amplifying the already-colder climates of the period (see figure 3.1). Positive feedback in this case is a consequence of ice sheet albedo, ice sheet elevation, and surface smoothness of the ice sheets. Because of their importance to the RCCE phenomenon, let's take a closer look at each of these characteristics in turn:

Ice sheet albedo. Albedo is a measure of how much of the incoming solar radiation is reflected. An absolutely black surface, for example, would absorb 100 percent, while an absolutely white surface would reflect 100 percent. Ice sheets are so white that they reflect considerable amounts of incoming solar radiation—up to 100 percent—and act like enormous refrigerators, cooling the air.

Ice sheet elevation. As one travels up in the lower few miles of the atmosphere (troposphere), temperatures drop. The central portion of the ice sheets covering North America and Europe were probably close to 2 miles high. Cold air is dense, which, along with gravity, causes downward movement of the air, in the case of ice sheets toward their edges and the coast.

Surface smoothness of ice sheets. The flow of cold, dense air off ice sheets creates strong winds, intensifying the strength of atmospheric circulation patterns. Wind strength and the overall vigor of atmospheric circulation is further reinforced by the fact that ice sheet surfaces are relatively smooth, compared with mountains, valleys, and oceans, allowing the strong winds to slide off the ice sheet and onto adjacent areas without losing speed.

The Younger Dryas as a Classic Example of a RCCE

Several massive RCCEs occurred throughout the glacial period. During some of these, temperatures dropped almost 54 degrees Fahrenheit (30 degrees Celsius) in central Greenland (see figure 3.2). The climate change event that first attracted attention from scientists is the Younger Dryas (YD). The YD was the most significant rapid climate change event that occurred during the last deglaciation (a period of removal of the ice sheets) of the North Atlantic region. The main phase of the deglaciation lasted from approximately 16,000 years ago to about 11,500 years ago (see figure 3.1). As seen in the GISP2 record a dramatic shift in temperature (based on our measurements of stable isotopes of water) and atmospheric circulation strength, or windiness (based on calcium), occurred about 14,500 years ago. In a matter of a few years, sea ice over much of the North Atlantic decreased significantly.

The YD event has attracted considerable attention for many years. It is recorded in a variety of paleoclimate media and makes a good case study for RCCEs in general. For example, the former extent of YD glaciers, which advanced at this time, is marked by moraines (debris deposited along former ice margins), in marine and lake sediments, and even in the remains of cold-loving plants and animals, such as beetles. Previous ice core studies focused on the abrupt termination of this event (without necessarily realizing that this and other RCCEs had very rapid

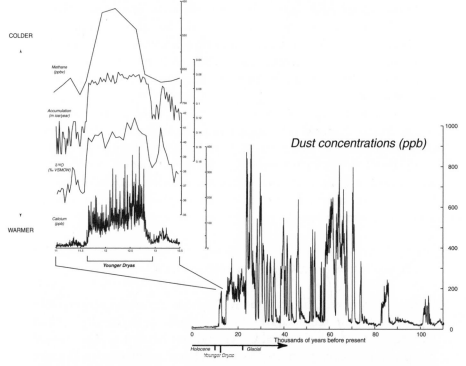

FIGURE 3.5. The Younger Dryas was an abrupt return to near-glacial conditions (15 degrees Celsius +/- 3 degrees or 27 degrees Fahrenheit +/- 5.4 degrees cooling relative to today), decreased accumulation rate, decreased CH4, and increased atmospheric dust, which lasted approximately 1,300 years and punctuated the transition from glacial to interglacial climates.

The 110,000-year-long calcium record from GISP2 ice core indicates the relative amount of dust in the atmosphere over Greenland. It documents not only the Younger Dryas but also many other abrupt, frequent, and massive changes in climate that characterize the glacial portion of the ice core record. *Figure at left modified from Alley et al., 1993; Grootes et al., 1993; Brook et al., 1996; Mayewski et al., 1997. Figure at right modified from Mayewski et al., 1994; 1997.*

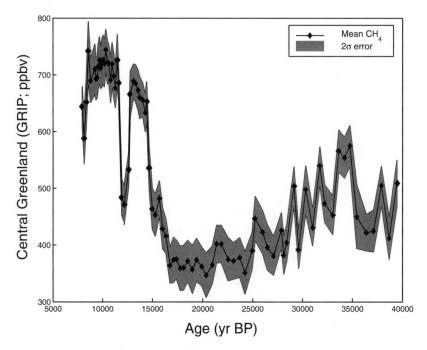

FIGURE 3.6. Methane record from the GRIP ice core from the period 5,000 to 40,000 years ago with standard analytical error. Levels of the greenhouse gas methane (CH_4) increase when biological material on land is in a moist environment. Therefore, increased levels of methane imply the presence of standing water rather than cold, ice-covered, or arid conditions. Levels of methane could only have increased in regions not covered by ice sheets during the glacial period (11,500 to 110,000 years ago). The rapid changes in methane seen during this time demonstrate that these events were not restricted to Greenland but were also found in temperate and tropical regions. This data set demonstrated that the search for rapid climate change events should definitely be extended outside of Greenland. *Data from Chappallaz et al., 1993.*

beginnings and terminations), because this transition marks the end of the last major climate reorganization during the deglaciation. The YD was dated with the GISP2 ice core (Alley et al., 1993) using precision measurements as a 1,300+/−70-year-duration event that terminated abruptly, as evidenced by a rise in temperature (Grootes et al., 1993), an increase in accumulation rate at about 11,640 years ago (Meese et al., 1994), and a decrease in continental source dust (Mayewski et al., 1993).

We used high-resolution (with a mean 3.48 years/sample) continuous measurements of GISP2 chemical species to reconstruct the paleo-environment during the YD because these series record the history of the major soluble constituents transported in the atmosphere and deposited

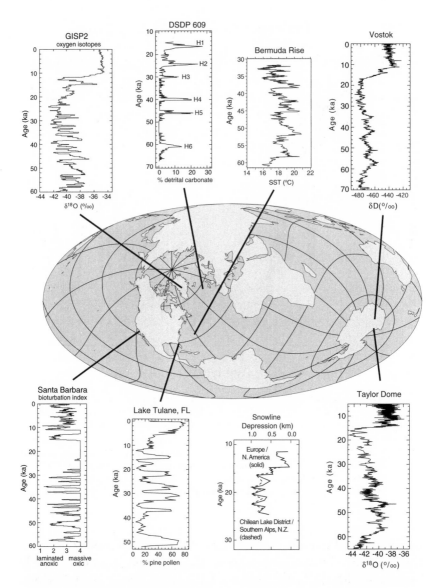

FIGURE 3.7. Examples of rapid climate change events found in a variety of paleoclimate records from throughout the world. The exact timing of these events from site to site is not yet sufficiently well dated to determine if all occur at the same time, in a phased sequence from site to site, or all at different times. However, the fact remains that natural climate variability is a critical component of climate over long periods of time and at a global scale.

Records displayed in this figure have been modified from the original by Kirk Maasch as follows: GISP2 oxygen isotopes (more negative δ¹⁸O colder) modified from Stuiver et al., 1995 and Grootes and Stuiver, 1997; Deep Sea Drilling Project (DSDP) record 609 from a North Atlantic marine sediment displaying periods of massive iceberg discharge that carries detrital carbonate into the North Atlantic defined by H (Heinrich) events modified from Bond et al., 1992; Reconstructed sea surface temperature developed from marine sediments recovered from the Bermuda Rise from Sachs and Lehman 1999; Vostok ice core deuterium isotopes (more negative δD colder) modified from Lorius et al., 1985 and Jouzel et al., 1987;

over central Greenland (see fig. 3.5). The calcium series provides a dramatic example covering the last ~10,000 to 18,000 years. We can see prominent periods of increased dustiness in this record, peaking approximately every 400 to 500 years and demonstrating the variability inherent in RCCEs.

Of greatest significance is the rapidity with which the YD and other RCCEs start and end. It is remarkable to see from figure 3.5 that the YD began in a decade or less and ended in a decade or less. The end of the YD also marks the official entrance into the Holocene, which is our our current interglacial climate period.

If we only found a few cases of rapid climate change events such as the Younger Dryas, and if they only showed up in the Greenland cores, these might still be dismissed as anomalies unworthy of too much attention. However, they are in fact found in many different records.

The climate change that accompanied the YD, for example, was not restricted to Greenland. The record of variations in the CH_4 concentration of trapped gases in the GRIP core (see fig. 3.6) shows that tropical and subtropical climates were colder and drier during the YD and during earlier RCCEs (Chappellaz et al., 1993). The major natural source region of CH_4 is low-latitude wetlands; higher atmospheric concentrations are due to the greater areal extent of tropical and subtropical wetlands. This finding suggested that we should look for similar events in other regions, because it appeared that the source of the CH_4 in the last glacial period may have been the lower and middle latitudes (Chappellaz et al., 1993). In addition to the RCCEs found in Greenland, they have also been discovered in a variety of other records (see fig. 3.7).

Multiple Controls on Climate

The debate over issues such as global warming often breaks down into a discussion of what appear to be polar opposites. Those who favor explanations based on anthropogenic causes point to the increases in greenhouse gases as the culprit, while those who favor natural causes point to the effects of changes in solar variability, volcanic effects, and other natural causes.

Santa Barbara marine sediments (more laminated equals colder) modified from Behl and Kennett 1996; Lake Tulane, Florida lake sediment (less percentage of pine equals colder) modified from Grimm et al., 1993; Snowline depression estimates from Europe, North America, South America, and New Zealand (greater snowline depression colder) modified from Denton, in press; Taylor Dome ice core oxygen isotopes (more negative $\delta^{18}O$ colder) modified from Grootes et al., 1994.

Scientists familiar with the field tend instead to think of *multiple controls* on climate (multiple forcing) rather than one control, and the GISP2 data is rich enough that we can finally consider creating the view of a series of multiple forcings on climate that includes most known factors, weighting them in terms of their overall effect. In retrospect, this of course makes a lot of sense. The Earth is a complex system, and its climate is a complex subsystem within it. No single factor would be likely to account for the behavior of the climate as a whole, or over long periods of time.

For example, much of what we found in the GISP2 record can be explained by changes in the amount of energy we receive from the sun, which varies depending on the Earth's distance from the sun, changes in the energy output of the sun, changes in the size and cover of ice sheets and sea ice, and changes in ocean circulation.

A lot of the credit for our knowledge of climate controls inferred from the GISP2 record goes to David Meeker, a mathematician at the University of New Hampshire, and a close friend of mine. With considerable multidisciplinary experience in the use of mathematics related to cancer research, sociology, and now climate change, he has brought important new insights to the work of climate researchers. Dave is an amazing combination of applied mathematician, cowboy (he was born and raised on a ranch in eastern Oregon), former bucking bronco rodeo competitor, and skilled New England sailor. In fact, some of our best scientific discussions have occurred on his sailboat out in the Gulf of Maine.

Based on Dave's work and the efforts of many other researchers, we now feel confident that rapid climate change events may be set into motion by the impact of several climate controls operating together, which sometimes create additive effects and sometimes cancel out each other's effects. This reasoning is important to our understanding not only of paleoclimate, but also of modern climate. After all, human effects on climate can be see as additional controls, but depending upon the timing and magnitude of natural controls, the human effects could be reinforced or reduced.

As it turns out, during most of the 110,000 years of the GISP2 record, all of the chemical species have fairly similar behavior when we look primarily at time scales of several decades. Higher-resolution views include more differences, which we discuss in chapter 5. All of the chemical species that function as a monitor for changes on land (the dust species: calcium, magnesium, and potassium), and all of those reflecting change over the ocean (sodium and chloride) exhibit similar behavior. When concentrations of all of these chemical species rise or fall, the atmospheric circulation system is more or less energized (wind speeds increase or decrease). For the atmospheric circulation systems over both land and ocean to intensify so greatly (as indicated by order of magnitude increases in dust and sea salt) during RCCEs we assume that most or all of the high-latitude (polar) atmospheric circulation system is energized during RCCEs. Increased levels of dust and sea salt indicate an increase in what we term the Polar Circulation Index (PCI) (Mayewski et al., 1997; Meeker et al., 1997). The PCI approximates more than 90 percent of the behavior (ups and downs) of the sea salt and dust. With this new approximation for polar atmospheric circulation, we investigated whether we could see any predictable patterns. In fact, we found several, which we describe below. Each of these could be approximated by a signal with a repeating cycle (period) represented statistically by bandpass components (these allow us some flexibility plus or minus 10 percent of the period). The periods identified in our record are unmistakable—they are significant at 99.9 percent (this means that there is a 0.1 percent possibility that the cycle occurs by chance).

We were encouraged to find so much recurring or periodic behavior in the last 110,000 years of climate history. The more order we find in nature, the better our chances of predicting future behavior, in our case, that part of the climate system controlled by natural factors. Mathematical analysis of the GISP2 record thus far reveals quite regular behavior on the order of 500-year cycles and longer. It seems that shorter period cycles do not always recur quite so regularly. We clearly need to refine this data to predict climate on time scales that will be of practical value to humans. In later chapters, we will discuss how we are trying to pull out even more signals from the record.

We identified eight major cycles in the PCI that together approximate close to 90 percent of the original record (another way of saying that the original record is very predictable). Do these cycles have any relevance in nature or are they just statistical trickery? As it turns out, they are extremely relevant. Let's take a look at why this is so.

The first four cycles (>70,000, 38,500, 22,500, and 11,100 years) identified in our record are all related to the amount of energy received on Earth from the sun (see figure 3.8). They have all been seen in other paleoclimate records, notably those developed from marine sediments,

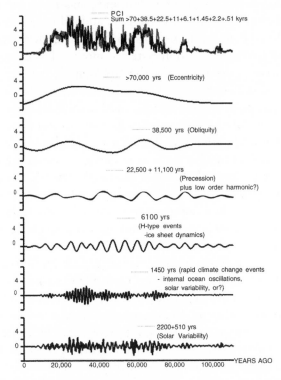

Atmospheric Circulation Forcing Frequencies
(expressed as bandpass components of the
Polar Circulation Index (PCI))

PCI
Sum >70+38.5+22.5+11+6.1+1.45+2.2+.51 kyrs

>70,000 yrs (Eccentricity)

38,500 yrs (Obliquity)

22,500 + 11,100 yrs
(Precession)
plus low order harmonic?)

6100 yrs
(H-type events
-ice sheet dynamics)

1450 yrs (rapid climate change events
- internal ocean oscillations,
solar variability, or?)

2200+510 yrs
(Solar Variability)

YEARS AGO

0   20,000   40,000   60,000   80,000   100,000

FIGURE 3.8. The PCI (Polar Circulation Index), top of figure, is a record developed from the analysis of GISP2 major chemistry (sodium, magnesium, calcium, potassium, ammonium, chloride, sulfate, and nitrate—together these represent more than 95 percent of all of the chemistry dissolved in the atmosphere). The PCI describes the dynamics (i.e., increase and decrease from mean values) of the polar atmosphere that envelops the source regions for chemicals transported to Greenland during the last glacial period (11,500 to 110,000 years ago). Thus the PCI provides a relative measure of the average size and intensity of polar atmospheric circulation. In general terms, PCI values increase (e.g., more continental dusts and marine contributions) during colder portions of the glacial period (stadials) and decrease during warmer periods (interstadials and interglacials) (Mayewski et al., 1997).

The PCI record is contrasted with the sum of the cycles estimated from this series. The sum represents about 90 percent of the signal behavior in the original PCI series, demonstrating that the original record is almost approximated by these cycles and that the record is comprised of a great deal of predictability. Statistical tests demonstrate that these cycles are more than 99 percent likely.

Major cycles derived from the PCI series include those with periodicities close to Earth's orbital cycles of eccentricity, obliquity, and precession (see figures 3.9 and 3.10). Other cycles are believed to be related to ice sheet dynamics (see figure 3.11), ocean circulation (see figure 3.13), and solar variability.

The 6,100-year cycle approximates the timing of Heinrich events—massive discharges of icebergs that carried glacially carved debris into the ocean (see figure 3.11). The 1,450-year cycle approximates the timing of many of the rapid climate change events in the GISP2 record. *Figure modified from Mayewski et al., 1997.*

which reveal change in climate through change in taxa, stable isotopes of marine shells, or changes in ocean chemistry (Shackleton and Opdyke, 1976; Hays et al., 1976; Imbrie and Imbrie, 1979). These cycles are related to the astronomical theory of climate change.

≡ **THE ASTRONOMICAL THEORY OF CLIMATE CHANGE**

In the mid-1800s, the Scotsman James Croll published an astronomical theory linking ice ages with periodic changes in the Earth's orbit around the sun. By the early 1900s, the Yugoslavian mathematician Milutin Milankovitch had refined and expanded this theory, offering three elements of orbital geometry that led to changes in the distribution and magnitude of incoming solar radiation, measured

FIGURE 3.9. The Yugoslav astronomer Milutin Milankovitch developed the astronomical theory of the ice ages in the 1920s and 1930s, shown here in figures 3.9a, 3.9b, and 3.9c.

Precession

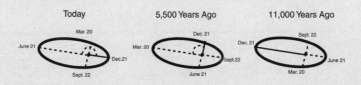

3.9a: Every 23,000 years, the Earth is closest to the Sun in an opposing hemisphere, resulting in the precession of the equinoxes. Viewing the cycle of this process, we see that the Earth was closest to the Sun during the June solstice 11,000 years ago and today is closest during the December solstice. For the Northern Hemisphere, this means that summers were warmer 11,000 years ago than they are today.

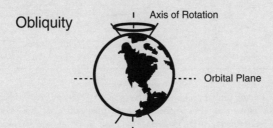

Obliquity

3.9b: The tilt of the Earth's axis of rotation in relation to the plane of its orbit varies between 21.8° and 24.4° over a period close to 41,000 years.

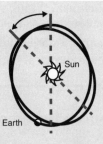

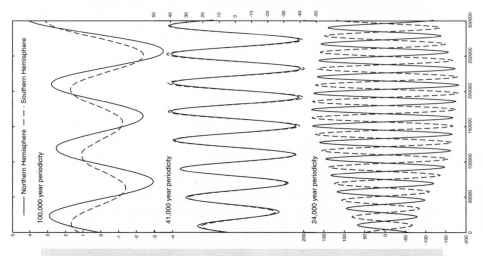

3.9c: The shape of the Earth's orbit around the Sun varies from near-circular to elliptical over a period of close to 100,000 years.

commonly in watts per square meter (for each 100 watts per square meter, envision a 100-watt light bulb held over a one-meter-square section of the Earth's surface). The orbital elements include eccentricity, obliquity or tilt, and precession of the equinoxes. Milankovitch's original theoretical calculations were modified by Andre Berger of the Netherlands (Berger, 1978) affording us theoretical views of changes in incoming solar radiation backward and forward through time.

FIGURE 3.10. Earth's orbital cycles for the last 300,000 years at two different latitudes (N60 and S60), plotted in terms of the amount of solar radiation (insolation) reaching these latitudes (watts/meter^2/year). Differences in insolation over time and over the planet are a major driving force behind climate change.

Why are the cycles seen in figure 3.8 related to Earth's orbital cycles? These cycles represent changes in our polar atmospheric circulation index (PCI). For example, when ice sheets in the Northern Hemisphere expand, the polar atmospheric circulation system expands and is energized (that is, wind speed increases). We have demonstrated the fact that our PCI is related to change in ice sheet volume by plotting the sea level record covering the last glacial/interglacial cycle compared to our PCI record. Michael Prentice, a colleague at the University of New Hampshire who studies sea level, compared the periods in our PCI record that relate to Earth's orbital cycles and found a close association in that sea level falls when the PCI falls (Mayewski et al., 1997). Why? It is because sea level is largely controlled by how much water is tied up in ice sheets, with lower sea level meaning larger ice sheets.

Why do the PCI cycles differ slightly from the calculated orbital cycles? Let's consider the following: We find a >70,000-year cycle related to an orbital eccentricity that is a 100,000-year cycle. Our record extends only 110,000 years back, so we cannot really see cycles as long as 100,000 years (we would ideally need a 200,000 to 300,000-year-plus record). We find a 38,500-year cycle that is very close to the 41,000-year obliquity cycle, and we find a 22,500-year cycle that is very close to the 23,000-year precession cycle. We believe that the difference between the PCI cycle and orbital cycle reflects the lag time that it takes the ice sheets to adjust to changes in incoming solar radiation. In fact, the PCI cycles usually start about 6,000 years after the change in incoming radiation changes. Finally, what of the 11,500-year cycle? It has been seen before in paleoclimate records and attributed to "ringing" in the climate system. We believe that if you hit the Earth with a change in incoming solar radiation, the effects may continue for a while, possibly yielding half-cycles of the original "hit" as the response gradually dampens (Mayewski et al., 1997).

The 6,000-year cycle in the PCI comes very close to approximating the timing of a series of very interesting events seen in the marine record of the North Atlantic. Called Heinrich events after their discoverer, these events are massive discharges of icebergs, which we can detect in marine sediment records as increased amounts of coarse debris that would not be found very far out into the ocean unless they had been "rafted" by icebergs (Bond et al., 1992). Ice sheets can only grow so large because they are limited in extent by their physical borders, such as oceans and mountains. In the case of oceans, the ice edge can extend

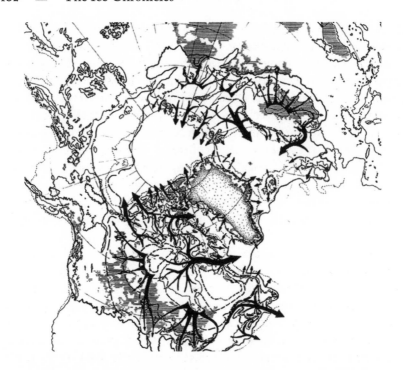

FIGURE 3.11. The ice sheets of the Northern Hemisphere at their maximum extent (18,000 to 23,000 years ago). Dark arrowed lines indicate regions of fast-moving ice called ice streams. These are the primary regions along which ice is transported to the edge of the ice sheet. Where these ice streams enter the ocean, they spawn icebergs. Massive increases in icebergs can result in the transport of large amounts of glacially eroded debris far out into the ocean because the debris is ice rafted (that is, rocks that would normally fall into the ocean off an ice edge are floated out to sea on icebergs). *Figure modified from Hughes and Denton (1985).*

only so far into the water before it begins to float. Discharge most likely occurs at the edge of ice streams, or regions of fast-moving ice (see fig. 3.11). If orbital cycles control the long-term size of ice sheets (as we believe they do by association with the sea-level record), then it must take time for the changes in incoming radiation to cause resulting changes in ice volume. Ice volume shifts are produced by variations in snow accumulation (warmer regions, more open ocean, more snow) and in temperature (warmer ice deforms more easily than cold ice). The so-called "response time" for this process appears to be 6,000 years, based on our record, meaning that ice was discharged every 6,000 years when ice sheets covered much of the high latitudes of the Northern Hemisphere (Mayewski et al., 1997). Analogs for this iceberg discharge theory come from modern observations of the edge of the Antarctic ice sheet, which responds to past and present changes in climate (see fig. 3.12).

What happened to the icebergs when they discharged into the ocean? In the case of the large discharges recorded by the Heinrich events, they probably covered much of the North Atlantic. The cooling effect would have increased sea ice extent in the North Atlantic, cooled air temperatures by association, and capped the surface ocean off from the atmosphere.

The next cyclical period identified in the PCI record operates at close to 1,450 years. If you look closely at figure 3.8, you will see that this period explains the timing of the RCCEs. When we first discovered this period in the PCI record, it was hard to imagine why RCCEs would be so regular. Such a period had not yet been found in other paleoclimate records and it was therefore hard to credit based on one record. It has now been seen in a variety of paleoclimate records, so we believe the cycle to be real, but we remain unsure of what causes it.

FIGURE 3.12. Antarctic icebergs produced as blocks break off from the large ice shelves that drain the Antarctic ice sheet. Picture taken from a C-130 aircraft viewing iceberg production off the edge of the Ross Ice Shelf. Mount Erebus, shown in the background, is the only active volcano on the continent. The icebergs in this picture are the size of city blocks. Several that are the size of Rhode Island have been observed over the years. *Photo by Paul Andrew Mayewski (1975).*

This is still one of the major mysteries involved in the study of ice age RCCEs, largely because we cannot tell the relative timing of these events from one site to another without extremely well-dated records, and few exist that are as well dated as the GISP2 and GRIP ice cores. The search is on for well-dated records from Antarctica leading to future generations of GISP2-like records, but that story is not yet complete. If we found that all of the 1,450-year RCCEs occurred at the same time all over the world, we would have to assume something like a change in output of the sun is the cause. If the events show a phasing sequence, operating first in the Antarctic, for example, and then traveling through to the Arctic, the cause would be likely to lie in the Antarctic.

We do not yet know the answer, but we do know that many marine sediment records documenting change in the distribution of both surface water and deep water show RCCEs, even if the timing relative to GISP2 is not definitive. If these occur in the ocean and in the atmosphere, what other causes might be at work other than solar variability? Is it possible that the ocean circulates with a period close to 1,450 years? So called, "simple deterministic feedback system" models that include changes in deep ocean temperature, sea ice edge extent, and carbon dioxide content of the atmosphere have resulted in climate oscillations that operate at about 1,300 years (Saltzman et al., 1981). This demonstrates that self-regulating aspects of the climate system may effectively feed off each other, and, unless these are significantly perturbed (by a new control, for example), may be set in a mode in which climate oscillates in a stable periodic way. Has such a mechanism caused the 1,450-year repeating RCCEs?

The answer may be found in any number of subtle factors, such as internal ocean oscillations. The oceans "turn over," producing a conveyor-belt phenomenon around the world, as shown in figure 3.13. For some reason, however, the conveyor belt shuts off repeatedly and regularly, and these shutdowns seem to be connected with RCCEs.

≡ THE OCEAN'S CONVEYOR BELT

Ocean water sinks when it is dense. The colder or the more saline (salty) the water, the denser it becomes. The North Atlantic is

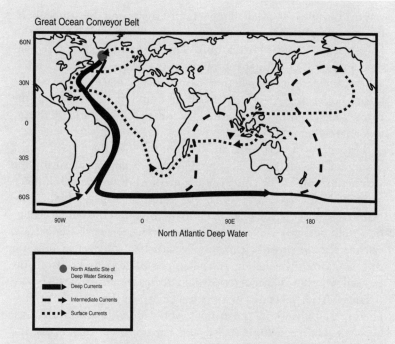

Great Ocean Conveyor Belt

North Atlantic Deep Water

- ● North Atlantic Site of Deep Water Sinking
- ▬▬▶ Deep Currents
- ‒ ‒ ▶ Intermediate Currents
- ■ ■ ■ ▶ Surface Currents

FIGURE 3.13. A diagrammatic view of the Great Ocean Conveyor Belt that circulates salt and heat throughout the oceans. Water sinks in the North Atlantic (and off the coast of the Antarctic—not pictured here), moves southward as North Atlantic Deep Water, and joins the Circumpolar Current that surrounds the Antarctic. Some returns to the North Atlantic Ocean as a Deep Current and the remainder rises to an intermediate depth, crosses the equator, and can rise with surface currents in the North Atlantic and other oceans. *Figure modified from matthias.tomczak@flinders.edu.au.*

unique because it is the only place in the Northern Hemisphere where significant amounts of dense water sink to great depths in the ocean. This water is not only cold (it is polar water) but it is also relatively saline. This is because winds blowing from east to west across the equator transport moisture from the Atlantic to the Pacific, leaving the Atlantic more saline. Waters that feed the Atlantic from the Mediterranean are also salty because they are nearly landlocked and moisture evaporates from the Mediterranean, leaving it saltier. This "conveyor belt" of dense water that is cold but mostly salty drives ocean circulation and, as a consequence, sends heat from the tropics up along the East Coast of North America, and then to the higher latitudes (see figure 3.13). The surface warm water that moves toward Europe is the well-known Gulf Stream and the atmospheric component of this system (the warm air over the Gulf Stream) is known as the "Nordic Heat Pump." This heat pump exerts a tremendous influence on climate throughout the entire Northern Hemisphere, and determines whether Europe, in particular, will be cold or warm. For example, evidence has been found of a shutdown in the 1970s, which may have accounted for some of the turbulence in European weather at that time.

The conveyor belt certainly did shut off with the advent of the Younger Dryas 12,800 years ago, based on examination of marine sediment records. The conveyor belt might slow down or shut down completely for several reasons. For example, icebergs (which are really freshwater ice formed on land) melting into the North Atlantic would decrease the salinity of the surface water of the oceans, causing the conveyor belt to stop because less-salty water does not sink so easily. Alternately, introductions of fresh water from increased river flow in the Arctic would decrease salinity in the North Atlantic. Also, decreases in moisture transport through the atmosphere east to west in the tropics would decrease the difference between salinity in the Atlantic and Pacific. In all such cases, the turn-over process would be slowed down or, in extreme instances, thwarted. Through any of these processes, heat transport to the high latitudes would be decreased, with dramatic effects within seasons to years, just as occurs in RCCEs. Note that not all of the causes for shutdown are related to the introduction of a cooler climate. For example, increased river flow in the Arctic would likely be a result of warmer temperatures. A climate surprise?

The 2,200-and 510-year cycles in the PCI are relatively small in magnitude compared with the previous cycles, but they are very intriguing because these periods are believed to be important in other environmental records (see figure 3.8).

A record of variation in carbon 14 (an isotope of carbon) found in the growth rings of trees may provide a clue to the influence of solar variability on climate. Changes in the output of energetic particles from the sun (called the solar wind) are believed to affect how much carbon 14 is produced in the upper atmosphere. We now think that high carbon 14-production is associated with periods of low solar activity (Stuiver et al.,

1991). Statistical analysis of carbon 14 collected from tree-ring records covering the last 10,000 years reveals cycles of approximately 2,200 to 2,400, 500, 200, 80 to 90, 22, and 11 years (e.g., Suess, 1980; Hood and Jirkowic, 1990; Sonett and Finney, 1990). Examination of the GISP2 polar circulation index (PCI) also reveals the same cycles, which may suggest a solar variability influence on this record. Examination of just the cycles close to 2,200 and 500 years in the carbon-14 tree-ring records and the PCI reveals that the cycles look very similar, and that, in both cases, they explain about 40 percent of the total signal in the last 10,000 years of the record (the Holocene) (Mayewski et al., 1997). Carbon-14 measurements are not available from tree rings prior to about 10,000 years ago, so we cannot test these series against the PCI record. However, we have learned some very important lessons. Solar variability is a critical control on the climate of the Holocene. The influence of solar variability on older climate is less clear because we do not have companion records for validation. New results from another chemical produced in association with solar variability, berylium 10, will soon shed light on the importance of solar variability during times prior to the Holocene.

Settling the question of solar variability's impact is a challenge, because the observational evidence for its importance is found in sunspot cycles (changes in the output of the sun's energy operating every 11 years). However, we only have measurements from the upper atmosphere (measurements from the lower atmosphere would be affected by clouds and the chemistry of the atmosphere) available from satellites for the past two to three solar cycles and the temperature changes during those cycles have been quite small. Current thinking suggests that simple changes in temperature produced by changes in radiation during solar cycles would not be enough to account for the dramatic behavior of the RCCEs. Something more must be going on to cause the shifts seen in the records.

One line of thought points to the fact that changes in the sun's output also transform the chemical composition of the atmosphere, including ozone abundance. This opens up a new view of climate change that implicates the role of human activity from a different angle. Humans, by increasing levels of tropospheric ozone through pollution, and decreasing levels of stratospheric ozone through the production of ozone-destroying substances such as chlorofluorocarbons, may be complicating climate behavior.

There is no verified (instrumentally observed) solar cycle that is longer than 11 years (because the data set is not long enough) that might show a greater influence on temperature. This means that for now the relationship between solar variability and the GISP2 record remains controversial.

Knowing whether a climate event occurs all over the planet at the same time would make a good case that solar variability, which is one of the few forces that affects the whole Earth, is a key factor, but it is not conclusive proof. The biggest challenge in determining the true cause or causes of the RCCEs is to obtain more records that are as well dated as the Greenland ice cores. Once such records are available, it will be possible for us to analyze whether the RCCEs occur at the same time all over the world, if there is a phasing in of the events by region, or if they are unrelated in timing.

Most of the evidence is pointing to RCCEs as global events, even if we are unsure whether the events all occur at the same time or not. Moreover, we are accumulating evidence that these climate change events are not only rapid, but also regular, and we are beginning to see the impact of more than one forcing factor in causing them to occur. We are therefore shifting our attention away from "warming" versus "cooling" to "stable" versus "unstable" as a view of climate. What RCCEs seem to be telling us is that this is a more realistic view, though it has not quite seeped into public consciousness as yet.

Moving Toward a More Complete Understanding of Climate Change

Examination of the periodic behavior embedded in the GISP2 record provides us with a new and more complete view of how climate operates. It is amazing that one single climate record can capture so much regularity in nature and so many controls on climate. However, the task is far from complete. This view does not show us clearly how the processes of climate forcing operate. In particular, we do not yet fully understand the non-linearities in the system. For example, does it take some critical number of years or critical amount of forcing before climate responds? Once climate responds, is the response simple or complex?

These are issues that scientists are still debating, using the GISP2 data as a foundation. For now, we can at least develop a running list (a precursor to an equation that describes full processes) of controls on

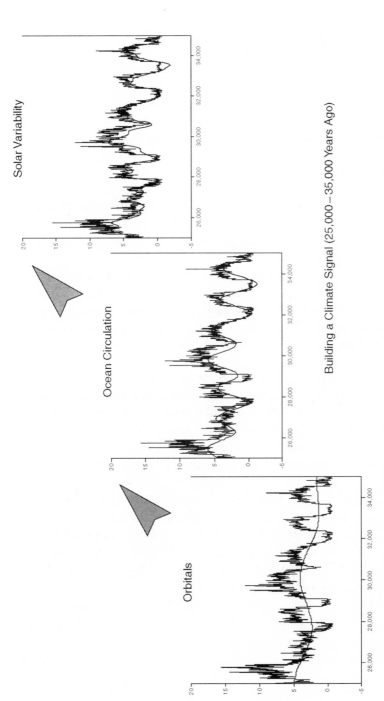

FIGURE 3.14. Climate is controlled by several factors. This figure reveals the effect of adding some of the climate change controls determined from the GISP2 record (see figure 3.8 for more details): the addition of Earth's orbital cycles plus cycles related to ocean circulation plus cycles potentially related to changes in the output of the Sun's energy (from Mayewski et al., 1997). Together these cycles closely approximate the behavior of atmospheric circulation (the polar circulation index, or PCI) over the last glacial period (11,500 to 110,000 year ago). The figure provides an expanded view of the PCI during the period 25,000 to 35,000 years ago.

climate. (See fig. 3.14.) We know from the 110,000-year GISP2 record that this list includes the following:

Earth's orbital cycles
+ oscillations in the behavior of ice sheets
+ oscillations in ocean circulation
+ changes in the output of the sun's energy

What else is involved? To the foregoing, we must add:

+ changes in greenhouse gas content of the atmosphere (greenhouse gases include water vapor in clouds, carbon dioxide, methane, nitrous oxide, and ozone—all of which are changed by natural and human activity)
+ changes in the concentration of dusts and acids in the atmosphere (produced by changes in moisture content on land, changes in wind speed over land and oceans, volcanoes, human activities, and biological productivity on land and in the oceans)
+ [additional variables as yet undefined]

We cannot yet put all of these controls on climate together to predict climate with any degree of precision. After all, purely random or chaotic behavior may also control climate. Pulling out the signal (the interpretable part of the record) is still a major challenge. Understanding climate has been a goal of humans for centuries, so we cannot expect to understand all of it in a few years. Nevertheless, we have made tremendous progress.

Understanding more about controls on climate and how climate has changed over time periods closer to the present means that we can now focus in the next chapter on the climate record of the Holocene (the last 10,000 or so years of Earth's history). It is within this time period that we can test the influence of climate on the emergence of civilization, and on the rise and fall of specific societies.

4

Climate Change and the Rise and Fall of Civilizations

Humans have occupied the planet for millions of years, and we know, from examination of marine sediments, that the 100,000-year sequence of ice sheet build-up and decay has been in action for at least one million years, if not longer. That means that, if natural climate alone is taken into account, we will very likely once again (in a few tens of thousands of years) have vast ice sheets return. There is also every reason to believe that the rapid climate change events that punctuate these 100,000-year cycles will continue. But what happens to the RCCEs when the ice sheets disappear and the Earth enters a milder period, like the current interglacial (the Holocene)? More important, perhaps, if rapid climate change events have occurred during the Holocene, have they had an effect on the development of civilization?

✳ CLIMBING GLACIERS BAREFOOT (WITH A BEAGLE)

I remember once, on an expedition to the Himalayas, waking up to the sound of our porters returning to the camp after an evening down the mountain with their families in the village below. I rolled over and looked at my wristwatch—5 A.M. These guys were amazing! They had left us late last night at 18,000 feet (5,486 meters), walked home 7,000 feet (2,132 meters) down the mountain, slept a few hours, and now they were back, ready to climb again. In the four and a half hours while they were at home, they had to eat as well, because this was during the Muslim holiday period of Ramadan, when devout Muslims eat only after sundown.

FIGURE 4.1. Ladahki porters and scientists climbing within one of the icefalls on the way up to the Nun Kun plateau (20,000 feet) in 1980. No photography or maps of this site were available when we went to Nun Kun, so the three icefalls we encountered offered a surprising and formidable obstacle. *Photo by Paul Andrew Mayewski (1980).*

As usual, they were barefoot, and they still carried their little beagle dog with them. Maybe it had been his barking that woke me up. Miracles of human adaptability to climatic variability, most of our porters steadfastly refused to use any footwear, even though I offered free boots to them. In fact, the boots we gave them were worth more as sale items than as shoes, and they put one porter through high school for a year.

The porters could climb up 500-foot near-vertical ice falls barefoot once we had set ropes for them, following us as we would laboriously set the ropes out while we climbed with our crampons on. I had to reset these ropes each day because the ice screws that held them would melt out in the noonday sun. Since the porters also wouldn't use any of the high-altitude climbing gear, they couldn't stay up on the mountain with us, and that's why they had to leave each evening. But they had tremendous endurance, and they needed very little sleep.

We were the first to do ice coring in the Ladakh region in the Indian Himalayas, but we couldn't have done it without our porters, who lived so easily in an environment that challenged us to the limit every day.

Rapid Climate Change Events (RCCEs) during the Holocene

To answer these questions about RCCEs and the Holocene, we turn to the Ice Chronicles and examine once again the last 110,000 years of climate record developed from the Greenland Ice Sheet Project Two (GISP2) ice core. Figure 4.2 offers plots of the change in concentration of continental source dust and sea salt in the GISP2 record. As noted in chapter 3, both dust and sea salt levels rise and fall together throughout much of the record. Increased levels of dust signal intensification of atmospheric transport from continental sources. During the coldest portions of the glacial age, the land surface was not as well stabilized by plant cover as during warmer periods, thus intensifying the availability of dusts. Because the dusts are derived from continents such as North

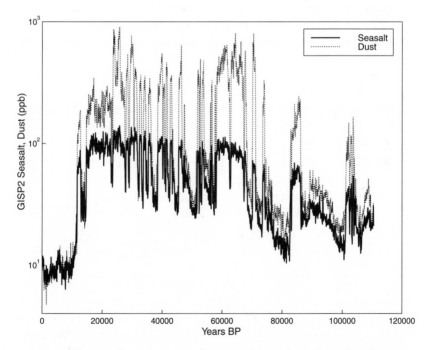

FIGURE 4.2. Change in the concentration of continental source dust and sea salt in the 110,000-year-long GISP2 record. A logarithmic scale is needed to see the small changes during the Holocene (last 11,500 years). Despite being small, these Holocene shifts in sea salt and dust demonstrate that the changes in atmospheric circulation that produced these shifts in sea salt and dust were sufficient to affect the course of civilization.

America and Asia, they are most likely carried to Greenland by west-to-east-traveling winds *(zonal winds)*. Increased levels of sea salt entering Greenland are most likely incorporated over the Atlantic, and therefore are carried south to north *(meridional winds)*. During RCCEs, both zonal and meridional winds intensify. Close examination of the Holocene portion of figure 4.2 demonstrates four major periods and several minor periods of intensified zonal and meridional winds since the end of the Younger Dryas (the last large RCCE before the beginning of the Holocene). Note that the scale for dust and sea salt during the Holocene is significantly different, with levels at one order of magnitude lower than during the glacial period.

Could these Holocene shifts in zonal and meridional winds be true RCCEs? There are two ways to determine the answer to this question, which is important because, if they are rapid climate change events, they occur during the period when civilization develops (the last 8,000 to 10,000 years) and might have had a detectable effect on the evolution of human society. This question is also important because such a finding would demonstrate that although the Holocene has a significantly milder climate than the glacial period, natural climate variability still plays a key role.

The first way to test the existence of RCCEs in the Holocene is to look for the same change characteristics as those found for the Younger Dryas and other RCCEs. RCCEs are present when we see intensified atmospheric circulation patterns, coupled with a decrease in temperature and snow accumulation rate in places such as Greenland, and a global decrease in the greenhouse gas methane (colder temperatures freeze standing water in which methane-emitting plants grow).

To search for these associations, we can look at the data on Holocene atmospheric circulation strength, represented in the graph in figure 4.3. The four largest changes in zonal and meridional atmospheric circulation are spaced at approximately 2,600-year intervals (O'Brien et al., 1995). This is a variation on the glacial age RCCEs, which were spaced at 1,500-year intervals. There may still be a 1,500-year pattern in the GISP2 Holocene record, but it is not as prominent as in the glacial age portion of the record. Interestingly, marine sediment records documenting iceberg discharge events in the North Atlantic during the Holocene reveal a pattern of nearly 1,500 years (Bond et al., 1997) that can be matched with a continental source—dust record for potassium from

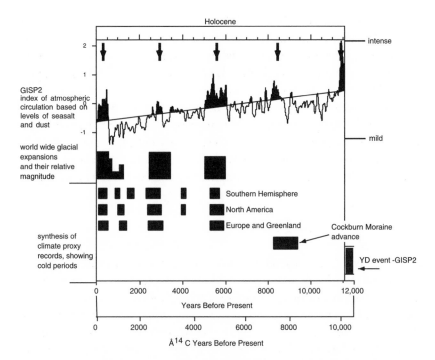

FIGURE 4.3. The Holocene (last 11,500 years) has been characterized by significant change in climate. GISP2 annually dated Holocene sea salt and dust combined to provide a record of atmospheric circulation (see figures 4.5 to 4.7 for calibration) that contains a quasi–2,600-year cycle (O'Brien et al., 1995), as does evidence for worldwide glacial expansions and their relative magnitude (Denton and Karlen, 1973) and a synthesis of various climate proxy records from Europe, Greenland, North America, and the Southern Hemisphere showing cold periods (Harvey, 1980), including the Cockburn Stade (Andrews and Ives, 1972; Alley et al., 1997; Stager and Mayewski, 1997), and the Younger Dryas (YD) event (Alley et al., 1993; Mayewski et al., 1993). Note prominent onset of the most recent period of intensified atmospheric circulation and cooling close to 600 years ago (A.D. 1400).

GISP2, which probably relates to changes in atmospheric circulation over Siberia and Asia (see chapter 5). It is encouraging to see the marine sediment record and ice core records reveal similarities as they did for the glacial portion of the record. For the oldest GISP2 Holocene large rapid climate change event, dating from 7,800 to 8,800 years ago, we see coincident decreases in accumulation rate, temperature (based on the stable isotope proxy), and methane, as expected from the glacial age characterization of RCCEs. Successively younger large rapid climate change events show similar associations, at 6,100 to 5,000, 3,100 to 2,400, and since 600 years ago.

Are they, then, still RCCEs? This question is made more difficult to answer because we know that the Holocene is complicated by changes in sea level, ice volume, solar variability, and human involvement. That might mean that glacial-age RCCE characteristics will not easily apply to the Holocene. After all, by 2,000 to 4,000 years ago, methane levels may have begun to be affected by human activities, instead of natural processes alone. By this time, a dramatic rise in agriculture had occurred, notably the cultivation of rice, which is a relatively large source of methane. The drop in methane that would be expected to occur in concert with a RCCE could have been cancelled by human involvement.

The Complexity of the Holocene Climate

Since the end of the Younger Dryas, the Northern Hemisphere has experienced significant environmental changes. As we can see in figure 4.4, ice sheet extent (ice volume) decreased rapidly from the end of the Younger Dryas to about 8,000 years ago, by which time ice extent in the Northern Hemisphere was fairly similar to what we see today. As the ice sheets melted, sea level rose from more than 131 feet (40 meters) below current sea level (approximately 11,500 years ago) to within 33 feet (10 meters) of present sea level by close to 8,000 years ago. Insolation (incoming solar radiation controlled by Earth's orbital cycles) changed dramatically. Approximately 11,500 years ago, winters were colder than today and summers were warmer. Thus, there was a greater difference between seasons than today.

Further, the Earth experienced several episodes when solar variability was particularly weak (shown as "T events" in figure 4.4). In addition, there were dramatic changes in flora, fauna, and the distribution of deserts. These all combine to make the Holocene a complicated period climatically.

George Denton, from the University of Maine, and Wibjorn Karlen, from the University of Stockholm, provided us with the second way to prove the existence of RCCEs in the Holocene. After having spent numerous field seasons in the Yukon Territory and Scandinavia mapping changes in the margin of Holocene age alpine (mountain) glaciers, Denton and Karlen wrote a paper summarizing their work and that of others. These glaciers are not large enough to dramatically affect sea level the way ice sheet ice volume changes can, but they are

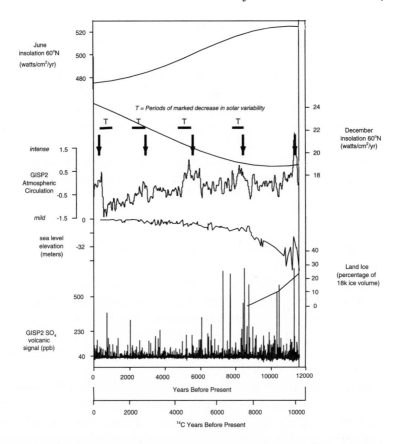

FIGURE 4.4. There are several causes for change in climate during the Holocene (last 11,500 years). Several direct and indirect causes that may affect the changes in atmospheric circulation revealed by the GISP2 record are captured in this figure.

Changes in insolation are produced by Earth's changing position in relation to the Sun. These changes in insolation differ according to time of the year (June and December values depicted) and latitude. As insolation changes, so does the distribution of heat over the surface of the Earth and as a consequence the position and the intensity of the storm systems that are recorded in the GISP2 record.

Periods during which solar energy output may have fluctuated coincide with the timing of increased storm activity revealed in the GISP2 record (see T's; after Stuiver and Braziunas, 1989).

Changes in land ice volume, which directly affects sea level, cause the albedo and geography of the Earth's surface to have changed dramatically over the Holocene, further complicating the study of Holocene climate change.

Changes in the timing and magnitude of volcanic events (Zielinski et al., 1997) can affect climate because volcanic materials (dust and sulfuric acid) shield the Earth from incoming solar radiation.

Piecing all of these potential causes together, developing records that allow characterization of other causes, and interpreting the resulting data is the key to understanding Holocene climate and our place in the Holocene climate system. *Figure modified from O'Brien et al., 1995.*

sensitive to much smaller changes in climate. The smaller a glacier, the less time it takes to respond to climate change. The "response time" for alpine glaciers is usually years to decades, while that for the glacial age ice sheets in the Northern Hemisphere or the Antarctic ice sheet is thousands of years. Their paper revealed globally distributed glacier fluctuations with a spacing at close to 2,600 years (Denton and Karlen, 1973). As figure 4.3 shows, the spacing coincides quite well with RCCEs in our Holocene GISP2 record. As a result, we can be sure that globally distributed cooling occurred, resulting in the advance of alpine glaciers that coincides with the timing of the largest GISP2 Holocene RCCEs.

Climate and the Development of Human Civilization

Having established the existence of Holocene era RCCEs, we can now ask whether they were severe enough to affect the course of human civilization. Tracking the human response to climate change in early times is a fascinating branch of climate research. It blends archaeology, history, anthropology, and mythology with paleoclimate research, meteorology, and other physical sciences.

Human involvement in climate change, and human response to it, is a familiar theme. In fact, an archaic example of a dramatic environmental change and its impact is among the most familiar stories in Western literature, that of Noah's flood. According to the Bible, Noah alone had access to the information that a tremendous storm would wipe out all life on Earth, and he was able to prepare himself and his family to survive it. The rest of humanity (staunch believers in gradual climate change!) went on with their lives as they had before, and all of them perished. Scholars and believers still debate whether the story of Noah is literally true, pure myth, or an amplification of an event that did occur, such as a localized flood. Regardless of the factual nature of the Noah saga, it is certainly an archetypical recounting of how a major environmental change might affect the world.

The national and international organization of our present civilization, with its advanced technology and sophisticated transportation systems, undoubtedly enables us, as never before, to rush help and supplies to relieve the immediate distress caused by natural disasters. Unlike Noah and his contemporaries, we also have the ability to forecast many of these situations before they take place, and people tend to believe what they are told by the media, so they prepare themselves as best they can. However, this complex worldwide community, with its interlocking arrangements and finely adjusted balances, may not be any more able than its predecessors to absorb the effects of major shifts in climate—particularly if they come on rapidly.

The run-up to the new millennium is instructive. Fear gripped much of the world because of concern, not about a natural disaster, but about a human-caused collapse of the world's information processing systems—the so-called Y2K bug. Computers, which have been widespread components of society for less than half a century, are now so integral to our lives that we cannot imagine managing our economic and social systems without them. With widespread interconnection of systems managing finances, power, and transportation, a small "glitch" can multiply its effects throughout the system, with disastrous consequences.

As it turned out, the system was more robust because of hard work than anticipated by Y2K worries. The glitches were few, and the inconveniences relatively trivial. However, the fears were real, and not altogether unfounded. We really do not know how well our society would cope with a major malfunction of the technology on which we now depend so greatly.

Just as it is useful to draw an accurate picture of natural climate to serve as a basis for the predictive models now under development, it may be valuable to look at cases of how earlier civilizations have responded to climate change as a basis for thinking about the most likely modern behaviors. For this purpose, the Holocene is an excellent laboratory. By studying this period, we may be able to gauge historical reaction to climate change. And perhaps the best place to look for human response to climate change might be the higher latitudes, where small changes in temperature can lead to large changes in sea ice extent, and to the mid to low latitudes where small changes in moisture can lead to severe drought or to islands being created where people are trapped by rising sea level or storms.

Let's examine several dramatic changes in environmental conditions that potentially affected the course of civilization. For each of these, the GISP2 Ice Chronicles and other paleoclimate records provide evidence of the climate change event. One of these events, the disappearance of the Norse colonies from Greenland, coincides with the onset of a major RCCE as defined already.

The others occur between RCCEs, suggesting that important changes in climate variability during the Holocene are not necessarily restricted to the major RCCEs seen in the GISP2 record. The archaeological record is complex, and by nature of its preservation will probably never be a continuous, annually dated record. Further, the archaeological record does not always come from places for which paleoclimate records are also available. Still, the association between archaeologists and paleoclimatologists is developing rapidly and offers exciting opportunities for better understanding the impact of climate on humans.

The Collapse of Civilization in Western Asia around 2200 B.C.

The role of climate in the development of human civilization has long been debated. One of the regions most intensely studied in attempting to understand this phenomenon is the Levantine Corridor of western Asia, an area stretching from Jericho to the Damascus Basin. The corridor has received so much attention because of the wealth of archaeological information in the region. For example, work by Ofer Bar-Yosef, an archaeologist from Harvard University, demonstrated that people in the Levant shifted to a lifestyle of cultivation and hunter-gathering some 10,300 to 11,600 years ago, just following the end of the Younger Dryas (Bar-Yosef, 1995).

This finding is extremely significant, because during the Younger Dryas (11,600 to 12,900 years ago) the region was very cold and dry, and populated primarily by hunter-gatherers. Then, between about 9,000 and 10,700 years ago, we find that Near Eastern agricultural communities had domesticated animals (Bar-Yosef, 1995). These findings demonstrate the potential effect produced by the last of the large RCCEs, the Younger Dryas. However, we still need to examine the more recent Holocene age RCCEs, and ask if they have had an impact on human societies.

Harvey Weiss of Yale University provides some answers with his findings concerning the collapse of the Mesopotamian Empire in 2200 B.C. (Weiss et al., 1993). A dramatic change in climate at that time, and others that are still being investigated, has now been located in the GISP2 records. The association between the GISP2 ice core record and the 2200 B.C. event is based on the fact that western Asia (notably Syria, which is modern-day Mesopotamia) receives its precipitation from

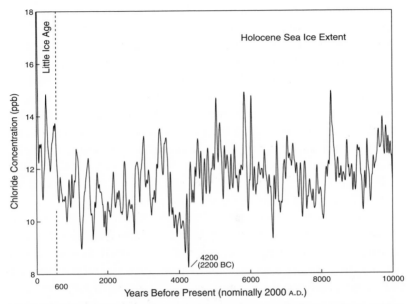

FIGURE 4.5. Change in sea ice extent in the North Atlantic over the last 10,000 years as determined from measurements of chloride in the GISP2 ice core. Chloride is transported as NaCl (sea salt) from the ocean to the GISP2 site. Increased levels characterize the Little Ice Age (note rise in chloride as of 600 years ago, A.D. 1400), a period with relatively greater storminess over the ocean and land plus more sea ice than during much of the Holocene. Conversely, decreased values are characteristic of periods with less sea ice and more summer-like conditions over the North Atlantic. The period around 2200 B.C. shows the most dramatic decrease in chloride in the record and hence more summer-like conditions over the North Atlantic. Around 2200 B.C. much of modern-day Syria and possibly regions as far away as eastern China experienced drought-like conditions. Syria is characterized by a summer dry period and a winter wet period. Increase in summer-like conditions could have heralded conditions conducive to a longer dry season during the 2200 B.C. collapse period, contributing to one of the more dramatic disruptions in ancient civilization, the collapse of the Akkadian Empire.

Interestingly, extremes in the GISP2 chloride record (interpreted as changes in sea ice extent in the North Atlantic) appear to coincide with at least two major events in the Holocene that had dramatic consequences on the course of civilization. These were the Little Ice Age (see chapter 5) and the collapse of the Akkadian Empire (2200 B.C.). While small compared to the glacial period rapid climate change events (11,500 to 110,000 years ago), Holocene climate change events appear to have been sufficiently large to affect humans.

westerly (west to east) winds that enter the region from the North At-lantic and Mediterranean during winter and spring. From late spring to early fall, the region is extremely dry because of a shift in the winds—winds blow over dry continental regions en route to this area at this time. This singular regional precipitation source and the timing of the precipitation, therefore, is responsible for both rain-fed agriculture of northern lowlands and plateaus as well as for the Tigris and Euphrates irrigation agriculture of the southern Mesopotamian plains.

How does the GISP2 record relate to conditions in western Asia? Dust and marine chemical species are carried to Greenland primarily during the winter/spring timeframe. Concentrations of both dust and marine chemistry vary during years when persistently winter/spring atmospheric circulation varies most dramatically over the North Atlantic. One of the lowest values of dust and the lowest of sea salt chloride in our record coincides with the 2200 to 1900 B.C. period. Thus, this period was characterized by lower persistence of winter/spring conditions over the North Atlantic. Low values of sea salt are also consistent with reduced sea ice extent (see fig. 4.5, studies in Antarctica and Greenland have demonstrated the association between reduced sea ice extent and reduced sea salt in ice cores). Less intense winter/spring atmospheric circulation implies less vigorous transport of moisture to western Asia and a lengthened dry season in that area. Of course, these conditions would have a major impact on agriculture and society as a whole.

Evidence corroborating the GISP2 assessment of a dry western Asia at about 2200 B.C. comes from sediment cores collected from the bottom of the Persian Gulf. They reveal that large amounts of dust blew off of the Mesopotamian land mass for several decades close to that time (Cullen et al., 2000).

Disappearance of the Norse Colonies in Greenland around A.D. 1400

As we have discussed elsewhere, according to medieval Icelandic sources, the Norse are believed to have arrived in Greenland close to A.D. 985, during a generally mild period called the Medieval Warm Period or MWP (see chapter 5). They settled on both the east and west coasts of Greenland. Researchers studying the archaeological remains of these colonies, including Tom McGovern from Hunter College, City University of New York, and Paul Buckland from University of Sheffield, England, have demonstrated through cultural remains and fossil beetle assemblages (beetle species are extremely sensitive to environmental conditions) that the Norse maintained a subsistence lifestyle based on milk and meat from domestic animals, which in turn made them largely dependent on hay crop production.

Thus, the Norse maintained a European culture quite unlike the indigenous Inuits' hunting and fishing existence. Norse clothing even adhered to European fashion of the period, while the Inuit wore clothing and made tools from locally abundant native sources. Under the mild conditions of the Medieval Warm Period, Norse life could in fact continue as it had in Europe. Yearly visits by European resupply vessels also guaranteed needed supplies to the Norse.

When we examine proxy temperature records from GISP2 produced as part of Lisa Barlow's Ph.D. dissertation work at the University of Colorado, and couple that with atmospheric circulation reconstructions based on chemical analyses from the ice core, we find that both the MWP and the RCCE that followed were climatically variable periods (figure 4.3, also known as the Little Ice Age; or LIA; chapter 5). The European culture of the Norse might have been greatly stressed at times during the cooler phases of the MWP, but by the rapid onset of the LIA (A.D. 1400 to 1420), cooler temperatures and the increased length of the sea ice season and extent of sea ice had eliminated any chance of resupply vessels from Europe getting to the colonies (see figure 4.5). To further exacerbate the situation, the growing season was shortened. The Norse in Greenland had disappeared by very early in the Little Ice Age, because they lived on the margin, geographically and in time, of a climate that did not allow them to subsist. By contrast, the earliest Eskimos had many centuries of climate change in which to adapt, and have therefore survived since their first entrance into the Arctic close to five thousand years ago.

Disappearance of the Maya Civilization A.D. 750 to 900

The Maya civilization developed around 3,000 years ago and collapsed by A.D. 750 to 900. Researchers have been able to correlate the most significant drying episode of the middle to late Holocene (A.D. 800 to 1000), developed from a lake sediment record in Mexico, with this dramatic and largely unexplained disappearance of the Maya civilization (Hodell et al., 1995).

The Mayans, who had created a society replete with grand cities, impressive pyramid-like structures, and advanced astronomical knowledge, seemingly just melted into the jungle almost overnight. We still don't know what happened to them, but it seems clear that climate change played an important role.

The GISP2 ammonium record indicates that the period of Mayan collapse was indeed a very dry period over all or much of the Northern Hemisphere. Ammonium in Greenland ice is derived from the emissions of biological material on land. Spikes in ammonium are believed to record forest fires, and background levels reflect the overall state of terrestrial biomass such as vegetation and organic rich soils. The dramatically low ammonium level that coincides with the collapse of the Maya civilization suggests that biomass levels were low, a finding consistent with the evidence of drought in the region at this time.

Descendants of the early Mayans live in Mexico and Central America to this day. Their presence is a reminder that unlike other so-called "lost civilizations," the Maya cultural achievements were not the stuff of myth and legend, but reflect historical reality. Could their demise represent a "rapid social change event" ("RSCE") that might appropriately be correlated with a rapid climate change event?

Lessons Learned from Past Civilizations

One of the most important tasks for paleoclimatologists is improving our understanding of Holocene climate, because it is within the Holocene that the boundary conditions for modern natural climate variability can be identified. It is also from this period that the relative importance of natural versus anthropogenic climate forcing can be assessed. Understanding modern climate and predicting future climate will require a detailed understanding of Holocene climate forcing and response.

In a *Time* magazine article, Eugene Linden wrote of the "warnings from the ice":

> The climate record shows that the whole 8,000-year span of human civilization, from the dawn of cities to space flight, has taken place during a period of extraordinary warmth and stability . . . The experience has left humanity with the notion that climate is warm and stable. But those who look at the past know different. "Climate is an angry beast," says [Columbia University's Lamont-Doherty Earth Observatory's] Wallace Broecker, "and we are poking at it with sticks." (*Time* magazine, 1997)

Of course, climate isn't really an "angry beast" that we can "poke at with sticks." It just seems that way sometimes. Climate is a complex system that we can understand or ignore at our peril, but Broecker's metaphor does bring home the importance of changing our perceptions of that system.

These realizations pose some troubling questions. For a long time, the history of human civilization has been written as if it were the result simply of human ingenuity, creativity, and a kind of inevitable march of progress. Few, if any, observers have attributed this dramatic change in the way people organize society and live together to the presence of a climate system.

Now that we have come so far, and suddenly know the truth about how we as a species got here, it's only natural to ask whether today's global society, representing the culmination of millennia of evolution, and nurtured by an anomalous climate system, can adjust to future climate events that will be unprecedented in terms of natural or human-influenced changes.

If we are going to survive, we first must recognize that the gradualist view of climate change carried with it certain assumptions about the current situation, and also had an impact on policy making. We have assumed, for example, that even if human activities forced the Earth to go through major shifts in climate patterns, society would have plenty of time to adjust to the new conditions.

However, our studies of the Greenland ice cores and other confirmatory data demonstrate that the conventional wisdom about climate change is wrong—GISP2 ice cores contain records of major changes in the climate system happening within the space of just a few years! If natural climate change in the past has reversed direction that quickly, then the time available to make adjustments—if any people had been around to make them—would have been quite small. If the time available for adaptation in the future is also short, the dislocations in society will be even larger. In other words, we may be standing on a kind of environmental precipice, about to step out into thin air without knowing it. Perhaps it is time to think of the climate system in a completely new way.

5

The Last Thousand Years of Climate Change

One of the reasons that scientists are so excited about having a 110,000-year-long climate record is that we can view our own climate in the twenty-first century along the same continuum as the climate that existed in prehistoric eras, in classical antiquity, in Napoleonic France, or the turn-of-the-century United States. This helps to make our work, which often seems to be esoteric and concerned with places and times that are distant from ordinary experience, interesting and relevant to others.

✳ PEAK EXPERIENCES

As I looked up at Mount Everest (Qomolongma in Chinese, Sagarmantha in Nepalese) rising dramatically over its neighbors in the Himalayas, I thought of those who had tried to conquer it (see figure 2.10). Everest is the highest mountain in the world at just over 29,000 feet (8,839 meters) above sea level. Quite by accident, we had just camped at about 22,500 feet (6,858 meters) above sea level on what we realized the following morning to be one of George Mallory's old campsites. He had made the first attempts to gain the summit in the 1920s, coming within a few hundred feet, but had disappeared on his last attempt with his climbing partner. Last year, Mallory's body was discovered, but the mystery surrounding what happened on his attempt at reaching the peak is still unresolved.

Sitting in camp and looking up at Everest, I also thought of the first people to have successfully climbed the summit and returned.

First conquered by Sir Edmund Hillary and his Sherpa guide Tenzing Norgay around the time of the International Geophysical Year and Sputnik's launching, the Everest summit remains a tantalizing challenge for climbers from around the world.

I felt close to Tenzing Norgay for many reasons. First, I have marveled for years at the strength and positive attitude of the Sherpa people, who have worked with us on expeditions in Nepal and Tibet. Second, I also had the pleasure of having Tenzing Norgay's son, Norbu, as my advisee while at the University of New Hampshire. He was a wonderful person, always reflecting those same qualities of grace and kindness of the Sherpa people, as if he were right at home in the New Hampshire hills rather than the mountains of Nepal. Tenzing Norbu combined his education and his experience in Asia to work in the adventure travel business.

Everest shows different faces to those who would come to know it well. When you approach it on the south side, from Kathmandu, coming up through the valleys of Nepal, Everest looks very much like one of many other Himalayan peaks.

From the north side, approaching over the Tibetan plateau, it's an entirely different story. Suddenly, the mountain looms up before you, 12,000 to 15,000 feet (3,657 to 4,572 meters) higher than anything around it, with the other peaks shrinking off to the side. From the plateau, you realize that you are already at 15,000 feet, but Everest is twice as high, and you can often see a banner of snow blowing off the summit.

Today, many of the people clambering up the Everest slopes are amateurs, taken to the top by professional climbers, and assisted by Norgay's successors, the modern Sherpa guides. These people are taking enormous risks. As you climb higher and higher, the oxygen saturation in your blood drops to 75 to 80 percent compared to the near 100 percent we enjoy at sea level, which dramatically decreases your energy and strength. The human body is really not designed to be at those high altitudes for very long. During the first days of our high-altitude expeditions to Asia, it is hard to stay focused and enthusiastic because of the oxygen deprivation. But as with any expedition, you learn what to expect and begin to draw yourself away from the negative aspects, trying hard "to be safe, have fun, and do our best job with the science"—the motto of my research team.

FIGURE 5.1. Mt. Everest from the south (Nepal) side (bottom) and from the north side (Tibet) (top).

When I say I'm going to Everest, some people think I'm planning to summit, but I never have and probably never will. I'm going for other reasons, because so many scientific opportunities are associated with the glaciers at these formidable elevations. Each year's snow is preserved to reveal new "ice chronicles" and the deposits marking former margins of the glaciers reveal changes in climate.

For example, ice cores that we have collected from the northern slopes of Everest tell us how the strength of the monsoon has varied over time. Monsoons are atmospheric circulation patterns that are characterized by dramatic seasonal shifts in wind direction. In the case of the Himalayas, the summer monsoon brings warm, moisture-bearing winds from the Indian Ocean and the Bay of Bengal as far north as the crest of the Himalayas and from there through high mountain passes to Everest. The winter monsoon in this region is characterized by dry, cold winds that blow southward from the Tibetan Plateau.

FIGURE 5.2. Most mountain glaciers on Earth have in general been retreating over the last few decades. In this figure, the retreat of the snout of the Far East Rongbuk Glacier is documented during the periods 1966, 1984, and 1997. The retreat is part of a pattern of general retreat for the region (Mayewski and Jeschke, 1979). The left lower corner of the photo displays the East Rongbuk Glacier, traveled on the ascent to the north side of Mt. Everest. The Far East Rongbuk Glacier was directly connected to the East Rongbuk prior to 1966. *Air photography provided by the Chinese Academy of Sciences (1966) and Bradford Washburn, Boston Museum of Science (1984), and from the Sino-American Everest Expedition (Qin Dahe and Paul Andrew Mayewski, 1997). Composite image by Jane Fithian, University of New Hampshire.*

Not every monsoon season is the same from year to year. More intense summer monsoons cause flooding from rain associated with the storms and later from melting snow. Less intense monsoons bring drought. By identifying patterns in this monsoonal system, we may be able to make better predictions about its behavior, allowing planning for drought or flood.

Our ice core research has revealed changes in the northward penetration of summer monsoonal air to the north side of Mount Everest. There has been a decrease in the last few decades, suggesting even greater potential for aridity over the already dramatically dry Tibetan Plateau. This recent change in climate is a potentially serious finding for inhabitants of the region. We don't yet know if this change is associated with warming in the lower atmosphere or is a normal part of the natural climate cycle.

Even the glaciers themselves in the Himalayas tell us a tale of recent and dramatic climate change. Former margins of the Far East Rongbuk Glacier, several kilometers north of the summit of Mount Everest, reveal dramatic retreat since the late 1960s. Along with this retreat has come considerable melting that is destroying what would have been as recently as the early 1970s or early 1980s an excellent site for the recovery of a well-preserved ice core record. The recent warming of the lower atmosphere appears to be destroying the very records that document it.

In this chapter, we will look at the most recent millennium of the Holocene, and discuss its importance in understanding the climate and weather that we ourselves experience every day. As it turns out, the Holocene contains a mystery concerning modern climate that examination of the Ice Chronicles has resolved—with startling results!

Of course, if you asked most people the question, "How do you like living in the Holocene?" you might have to duck—they would probably think you were insulting them, rather than talking about the climate! However, we are indeed living in the Holocene, and revelations from the Greenland GISP2 records have set off a spate of explorations in other parts of the world to learn more about it. The Greenland findings have encouraged us to look closely at the Holocene period because so many of our assumptions concerning this epoch were wrong. The new research has also told us that we need additional records from other parts of the world to extend and confirm what we discovered in Greenland.

The latest portion of the Holocene, in particular, is so important because it really is part of our own experience of climate and shapes our understanding of what we expect "normal" climate and weather should be. It also illuminates the story of how modern civilization has unfolded in that context.

"Normal" certainly belongs in quotation marks because climatologists, especially those specializing in paleoclimate, are beginning to view the Holocene in terms of the questions it raises more than the answers that it provides. As the answers do finally emerge, the past 1,000 years may or may not be characterized as normal. While relatively few people are aware of the connection, modern history is dated, by the Western world at least, with the onset of the late Holocene. It was also some 2,000 years ago, from the time of the birth of Christ, that we measure history as A.D. (Anno Domini) or C.E. (Common Era). Thus, the late Holocene marks the onset of modern history as well as modern climate.

When we look at Holocene data, two climate events leap out and demand our attention. They are known as the Little Ice Age, or LIA, and the Medieval Warm Period, or MWP. The MWP, which was relatively warm when compared with the LIA and perhaps even much of the twentieth century, started in the first few centuries of the first millennium A.D., included the Middle Ages, and ended around A.D. 1300 to 1,350. It was followed by the cooler LIA, a period that lasted from about A.D. 1400 to well into the early 1900s and perhaps longer.

During this time, the climate was generally cooler and glaciers were more expansive than either during the MWP or today. It was during the MWP that the Norse sailed to Greenland and established their colonies, which they abandoned as the North Atlantic grew colder during the ensuing Little Ice Age.

Characterization of climate during the MWP and LIA is possible only by examining many geographic regions and viewing climate change not just as a shift in temperature but also as change in precipitation and atmospheric circulation. This has been conclusively demonstrated in the fascinating compilations produced by the "Father of Paleoclimatology," Hubert H. Lamb, and by Jean M. Grove, author of *The Little Ice Age*. The evidence points convincingly to a dramatic difference in climate over much of the world for these two climate eras. For example, the ancient city of Petra, in what is today the desert region of Jordan, thrived from 300 B.C. to A.D. 100. Great cities flourished on the Turpan Desert in central Asia along the famous trade route known as the Silk Road, from A.D. 200 to A.D. 1000 during the MWP.

FIGURE 5.3. The once (300 B.C. to A.D. 100) thriving city of Petra is today in the middle of the Jordanian desert.

Warm weather in southern England during medieval times led to the appearance of vineyards, which then disappeared during the LIA. Tales of severe freezing on the Thames, starvation throughout Europe, and expanded range and duration of sea ice across the high latitudes characterized the LIA.

The Little Ice Age is especially interesting because it extends into a time that is not far from recent memory, when industrialization accelerated and human beings began adding a wide variety of pollutants to the atmosphere. Then, over the past century, temperatures rose above the typical values of the LIA and glaciers retreated around the world. This trend forces us to ask whether the temperature rise is part of a natural termination of the LIA, or if human activities have caused a non-natural end to the LIA. It's kind of a scientific mystery story, with a title like "Who killed the Little Ice Age?" or "Did the Little Ice Age die of natural causes?" or "Is the Little Ice Age really dead?"

The LIA is also extremely important in providing clues to modern climate behavior, because the GISP2 records establish that the LIA began more abruptly than any of the other RCCEs of the Holocene, as shown in chapter 4. Ice cores from the Andes and Antarctica confirm this finding.

In the words of Charles Officer and Jake Page (1993):

Early in the fifteenth century, around the time of Joan of Arc, the world began to turn cold. All over Europe, the relatively warm and moist climate of what is now called the Medieval Warm Period was replaced by temperatures that averaged 2.5 degrees Fahrenheit lower, a decrease in rainfall to about 90 percent of the previous average, and a greater variability in the seasonal weather from year to year.

Conventional wisdom holds that the Little Ice Age is now over, and indeed the average temperature did rise about 1 degree Fahrenheit in the Northern Hemisphere from 1900 to 1940 (see Technical Box to follow). In the Southern Hemisphere, a steady rise has been the norm, with an increase of about 1 degree Fahrenheit over the past century. But then an unexplained cooling period lasted from 1940 to 1970, followed by warming.

≡ **HOW COULD THE ATMOSPHERE COOL IF GREENHOUSE GASES ARE ON THE RISE?**

Although temperatures have risen in both hemispheres over the last century, there have been years and decades when the increase was interrupted by a period of cooling in the lower atmosphere. One of the most notable periods, and one that we have discussed earlier in this book, is the 1940 to 1970 cooling of the atmosphere, primarily over the North Atlantic. A simple comparison between the patterns of change in some primary parameters that control climate provides insight into the potential cause or causes of the 1940 to 1970 cooling.

Comparing the temperature trend for the Northern Hemisphere and one potential control on climate—the greenhouse gas carbon dioxide—demonstrates that while the overall increase in temperature on the scale of a century follows a trend similar to that in carbon dioxide, the 1940 to 1970 cooling is in fact not associated with any decrease in carbon dioxide. We therefore must look for different causes of the cooling.

Other potential causes for this phenomenon include the shielding of incoming solar radiation by increased levels of sulfuric acid in the atmosphere. Higher levels of sulfuric acid associated with volcanic emissions can certainly lead to cooling of the lower atmosphere as

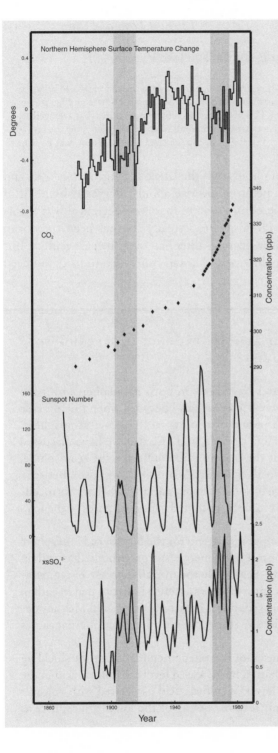

FIGURE 5.4. Mean annual surface temperature has risen globally over the last century as a consequence of multiple controls on climate. In this figure, change in Northern Hemisphere annual surface temperature (°C) (Hansen and Lebedeff, 1988) is seen to rise with the same overall trend as that produced by the anthropogenic rise in carbon dioxide, demonstrating the association between temperature and greenhouse gases. However, the overall rise in temperature is punctuated by periods of several years of decreased temperature during the early twentieth century and the 1960s to 1970s (e.g., shaded areas in figure). The periods of decreased temperature coincide with periods of decreased solar output (based on the sunspot number) and increased levels of sulfate. Notably, the increases in sulfate portrayed in this figure ($nssSO_4^{-2}$; Mayewski et al., 1986; 1990) are produced primarily as a byproduct of fossil fuel burning, although some are a consequence of volcanic activity (see figure 5.18). Further, sulfur emissions from volcanoes and fossil fuel burning reside in the atmosphere at most one to two years and more likely only several weeks and they usually have only a regional effect on climate, unlike greenhouse gases that are globally mixed. This information demonstrates the importance of appreciating both natural and anthropogenic controls on climate change.

much as 2 to 4 degrees Fahrenheit (1 to 2 degrees Celsius) for major eruptions such as the Tambora eruption of A.D. 1815 that produced the "year without a summer" in eastern North America and western Europe.

Benjamin Franklin was among the first observers to note the association between increased levels of volcanic dust and cooling. We now realize that in addition to cooling associated with dust, cooling is also associated with sulfur gases that combine with water in the atmosphere to form sulfuric acid. The sulfuric acid travels farther in the atmosphere than dust and often has a more far-reaching and longer-duration cooling effect. As we can see in the sulfuric acid (non–sea salt SO_4^{2-}) data (developed from an ice core recovered from southern Greenland) presented in figure 5.4, levels have risen dramatically since the beginning of the twentieth century in response to increased industrial activity and the burning of sulfur-rich fossil fuels.

Levels rose markedly from 1940 through the 1970s in response to industrial demand associated with World War II and the post–WW II industrial boom. Regions of particularly intense industrial source sulfur emissions rim the North Atlantic; hence the potential cooling in this region. This is clearly an instance in which human activity has affected the climate, though in the direction of regional cooling, rather than global warming.

In addition, portions of the cooling period noted in the figure were associated with a decrease in sunspots. Although the association between sunspot cycles and climate, such as the quasi-11-year cycle portrayed in this illustration, is not well understood, many highly significant associations have been noted between increased or decreased sunspot numbers and increased or decreased surface temperatures. We will examine these associations in more detail later in this chapter.

Perhaps the Little Ice Age has in fact been terminated naturally, and increasing CO_2 levels are accelerating a warming trend. But what if, as some researchers believe, *the Little Ice Age has not yet ended,* and the warming we are now seeing runs counter to the normal direction of natural climate? This means that the warming trend is not simply accelerating the direction in which natural climate is going, but is actually *reversing* it as a countervailing force.

If this is so, it means that the warming trend is even more significant than we had ever imagined, and requires further explanation. However, temperatures were not the only indicator of the LIA's onset. The appearance of the Little Ice Age was foreshadowed by globally distributed cooling, as indicated by the advance of mountain glaciers in Asia, North America, South America, Europe, New Zealand, and the polar regions, as shown in chapter 4 (figure 4.3). The Little Ice Age was also marked by a change in atmospheric circulation that produced increased storminess in at least the higher latitudes.

≡ STORMY WEATHER: THE MEDIEVAL WARM PERIOD AND LITTLE ICE AGE

The skills of Dave Meeker, the rodeo cowboy-turned mathematician, were certainly put to the test when we decided to compare all of the instrumental records of atmospheric pressure from the Northern Hemisphere to our ice core chemistry records. Atmospheric pressure is measured as force per unit area in millibars (mb), and the atmosphere surrounding Earth is known to exert a mean pressure of 1,013.25 mb at sea level. Departures from this sea level pressure (SLP) caused by differences in temperature, moisture content, dust content, and other factors result in areas of high and low pressure over the Earth's surface.

In combination with forces exerted by the Earth's rotation, winds in the Northern Hemisphere move in a counterclockwise rotation around low pressure areas and clockwise around highs. When these winds blow from the ocean toward an ice core site in central Greenland, we expect to find that the deeper (lower) the low pressure region, the stronger the transport of chemicals (for example, sea salts) from the ocean toward central Greenland. If a high pressure region is centered over land, then when pressures increase, there is intensified transport of land source chemistry (for example, dust). If this transport is in the direction of central Greenland, then more continental source dusts arrive at the ice core site.

We asked Dave Meeker to compare the 1,080 sites containing records of average monthly sea level pressure in the Northern Hemisphere for the period of collection of instrumental records (1899 to present) to our ice core record from GISP2. What we found was truly amazing. The results generated by this effort allow us to fore-

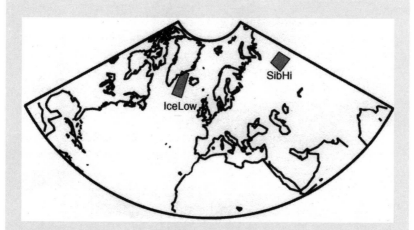

FIGURE 5.5. Ice core chemical records can be used to understand changes in the intensity and location of atmospheric circulation systems and through this association provide a means for understanding changes in atmospheric circulation prior to the very short period of instrumented climate. In order to develop a calibration between atmospheric circulation and ice core records, it is necessary to understand the modern distribution of sea level pressure (SLP) over the region that is potentially related to the ice core. Instrumental data is available for understanding the distribution of mean sea level pressure over the Northern Hemisphere for the period 1899 to present (Hansen and Lebedeff, 1988).

Sea salt (ss) sodium (Na) is transported between the ocean and the GISP2 site in winter, and non–sea salt (nss) potassium (K) is transported from the land to the GISP2 site in the spring. Changes in SLP (where the mean level at sea surface is 1,000 millibars) reveal changes in the strength of atmospheric circulation. During winter, a low-pressure field over the North Atlantic (Icelandic Low, labeled IceLow on the map) drives winds off the North Atlantic counterclockwise toward Greenland as SLP decreases. In the spring, a region of high pressure over the general area of Siberia (Siberian High or SibHi) drives winds toward Greenland in a clockwise direction as pressure increases. Analysis of the change in ssNa and nssK compared to changes in winter and spring SLP, respectively, over the Northern Hemisphere for the period of data overlap reveals very strong associations between ssNa and the Icelandic Low and nssK and the Siberian High (Meeker and Mayewski, 2001). Also see figure 5.6.

cast changes in sea level pressure back in time at least through portions of the Medieval Warm Period (MWP), showing how atmospheric circulation in the lower atmosphere has changed from the MWP to the Little Ice Age (LIA).

In the first set of figures, sites with the highest correlation between ion chemistry and SLP are revealed. The strongest associations were between sea salt sodium and the Icelandic Low (IceLow) and non–sea salt potassium (dust) and the Siberian High. This is based on a comparison of the data for the period 1899 to 1987, where 1899 represents the start of the instrumental record of SLP and 1987 the most recent ice core data point used in the comparison.

The *Icelandic Low* is the dominant winter surface atmospheric pattern over the North Atlantic. This low not only pushes oceanic air into central Greenland but is also the birthplace of the so-called "Nor'easters" that frequently hit the New England coast. Winter is the time of year when sea salt sodium is known to reach its maximum

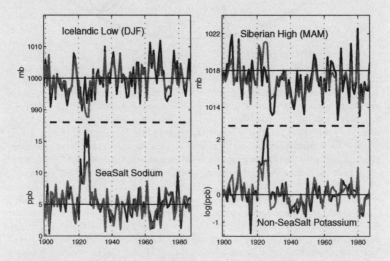

FIGURE 5.6. The association between GISP2 ssNa and the wintertime strength of the Icelandic Low whose winds blow sea salt (e.g., ssNa) is very close as demonstrated by the statistically derived relationship shown by the gray line in the upper left portion of the figure. The gray line closely approximates changes in Icelandic SLP. In the lower left graph, the same gray line approximation reveals a very close, but inverse association between Icelandic SLP and ssNa, such that decreased SLP (stronger winds generated toward Greenland) correlate with increased ssNa deposited at the GISP2 site.

The association between springtime strength of the Siberian High and nssK is also very close. Increased Siberian High SLP (stronger winds from Siberia to Greenland) is associated with increased deposition of nssK at the GISP2 site.

Note that while the changes in SLP associated with changes in ssNa and nssK appear small, up to 16 millibars for the Icelandic Low and up to 3 millibars for the Siberian High, compared to a mean SLP of 1,000 millibars, they are sufficiently large to cause major storms. *Figure modified from Meeker and Mayewski, 2001.*

in central Greenland snow and when marine storms are the most intense. Continental source potassium (that portion of the potassium not related to the oceans as a source) known as non–sea salt potassium, is most strongly associated with changes (increased levels of non–sea potassium are associated with increased SLP) over high pressure regions in Siberia during spring (March–April–May [MAM]). This high is called the *Siberian High* and spring is when most of the nssK is transported to central Greenland.

The next step in this comparison is to demonstrate whether the association between chemical concentration and SLP exists year to year over the full period of overlap between the data. To compare the association year to year, we employed a statistical tool that is used commonly in meteorology. The tool, an empirical orthogonal factor (EOF) analysis, looks for associations and generates an approximation to both records (chemistry and SLP). In the case of ssNa compared with the Icelandic Low and nssK with the Siberian High, we found matches that were close to 70 percent of the signal.

The fit is imperfect because not every fluctuation in chemical concentration is related to change in SLP. Other factors control chemical concentration at the drill site (e.g., moisture content of the air, topography, and temperature). The fit, however, is still very close, allowing chemical concentration to act as a proxy for the behavior of SLP back in time prior to actual measurement of SLP. With the chemical proxies for the Icelandic Low and the Siberian High, we can examine the behavior of these systems over the MWP and LIA. We found variability in both systems, but this variability is generally greater during the LIA. This suggests that storminess over the North Atlantic and Siberia was more variable and intense during the LIA. Most interestingly, modern-day levels of SLP have yet to return to those of the MWP. We take this as a strong indication that atmospheric circulation patterns established during the LIA continue today. Comparison between the proxies for the Icelandic Low and Siberian High versus ^{14}C residuals (a proxy for solar variability) demonstrate again the importance of solar output controls on climate.

Findings developed by one of my former Ph.D. students, Karl Kreutz, now on the faculty at the University of Maine, reveal a similar situation in West Antarctica. Using Dave Meeker's mathematical expertise as a guide, Kreutz found a proxy for the *Amundsen Sea Low,*

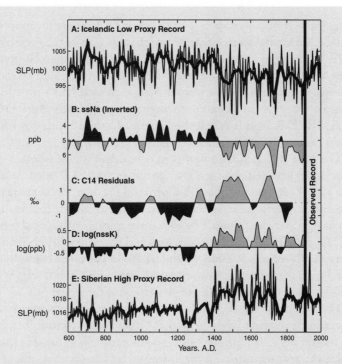

FIGURE 5.7. The observed or instrumental record of sea level pressure (SLP) for the Northern Hemisphere extends back only to 1899, but ice core records can be used to extend these records back in time significantly longer. In (A) on this figure, an approximation (proxy) record for 3-year averages of the Icelandic Low SLP extending back A.D. 600 is presented based on a 3-year smoothed GISP2 ssNa (B) series using the approximation presented in figure 5.6. Similarly, an approximation for the Siberian High is presented in (E) based on nssK data (D). For simplicity, the ssNa and nssK proxies for the Icelandic Low and Siberian High, respectively, are smoothed to 20 years. This smoothing reveals a clear sharp break in the style of atmospheric circulation affecting the North Atlantic and Eurasia as of A.D. 1400. The change is characterized by decreased Icelandic Low SLP and intensified wintertime circulation over the North Atlantic plus increased Siberian SLP and intensified springtime circulation over Eurasia. This increase in storminess as of A.D. 1400 heralds the beginning of the Little Ice Age in these regions (Mayewski et al., 1993a; O'Brien et al., 1995; Meeker and Mayewski, 2001). Prior to A.D. 600 and until A.D. 1400, during the Medieval Warm Period, it was much milder, in terms of storminess, in both the North Atlantic and Eurasia.

In (C) on this figure, an approximation for solar variability is provided, utilizing the ¹⁴C residual series developed by Stuiver and Braziunas (1993). High levels of ¹⁴C correlate closely with periods of low sunspot activity (low solar output). Note strong association between low values of ¹⁴C (high solar activity) and pre-A.D. 1400—Medieval Warm Period—atmospheric circulation and generally higher levels, and post-A.D. 1400—Little Ice Age—intensified atmospheric circulation. Note further the close association between particularly high levels of ¹⁴C (low solar output) and the most intense periods of atmospheric circulation during the Little Ice Age.

an Antarctic version of the Icelandic Low. This feature was weaker during the MWP and has not yet returned to pre-LIA conditions. Of further interest is the fact that the increase in ssNa at both Antarctica and Greenland sites comes very close to happening at the same time, the early 1400s.

This raises the question: Did the LIA start all over the world at the same time, or did the beginning differ in some regions, or not start at all in other regions? We do not yet know the answers to these questions, but once they are better understood they will help us to determine what caused the LIA. A globally synchronous start implies a very strong control factor, perhaps an extraterrestrial control such as change in energy output of the sun. On the other hand, if the start of the LIA turns out to be phased from site to site, we will want to go to the site where it first started to seek out the cause or causes.

One thing is, however, clear: The LIA is not dead!

While the relatively cooler temperatures of the LIA began to dissipate as early as the late nineteenth to early twentieth century, the atmospheric circulation patterns characterizing the continuing presence of the LIA have remained in place; there has yet to be a full return to the kinds of patterns that characterized the MWP. Moreover, if the LIA were truly over, it would not only be the most abrupt RCCE (Rapid Climate Change Event) in terms of how it began, it would also be one of the shortest of the last 110,000 years. Based on what we know of the GISP2 records, it would not be anomalous for the LIA to last another 200 to 500 years, because RCCEs typically last a thousand or more years, and the LIA only started 600 years ago.

Thus, researchers familiar with the Ice Chronicles and with RCCEs are forced to ask the question, "If the LIA isn't really over, why is global warming occurring, rather than continued cooling?" The debate is now in the public arena, with the media carrying numerous articles concerning whether warming seen in various parts of the globe is natural or not. For example, a story by Thomas Hayden and Sharon Begley, "Cold Comfort," appeared in *Newsweek* in 1997:

> Average annual air temperatures on the Antarctic Peninsula have climbed 5 degrees Fahrenheit over the last 50 years, 10 times faster than the global rate. Average midwinter temperatures there are up 9 degrees. The cause could be natural climate fluctuations. Or it could be global warming induced by the heat-trapping

"greenhouse" gases emitted into the atmosphere. According to computer models, man-made warming will be more extreme at the poles and show up there before it's detectable at mid latitudes. (Hayden and Begley, 1997)

Another statement of the controversy appeared in the Washington *Post* more recently: "Is the observed increase in the worldwide average temperature—around 1.1 degrees Fahrenheit over the past 100 years—genuinely abnormal, or is it well within the bounds of natural variability?" (Suplee, 2000).

Those supporting the position that the warming is natural argue, correctly, that RCCEs dramatically move climate in one direction or the other, but variability can still take place even within a RCCE. Thus, while the LIA might be propelling the climate toward colder temperatures, it is possible that we would still be able to observe warming periods within it. After all, we know that the LIA was, or is, a period of relatively cooler, but nevertheless variable, temperature. It is important to remember that variability in climate operates at many scales.

The climate was not consistently colder during the LIA or warmer during the MWP. However, the fact that atmospheric circulation patterns have not emerged from their LIA state of activity, while temperatures have risen, lends support to the idea that at least portions of the twentieth century warming are not caused by natural controls on climate.

☰ POLITICS, HISTORY, AND CLIMATE CHANGE

The debate over the significance of trends in temperature and the cause of such change is not a new one. In 1782, Thomas Jefferson wrote in "Notes on the State of Virginia" that "Snows are less frequent and less deep" than the elderly in the community had remembered from their earlier years. Noah Webster presented an innovative explanation for Jefferson's observations, stated in an address to the Connecticut Academy of Arts and Sciences in 1799:

> It appears that all the alterations in a country, in consequence of clearing and cultivation, result only in making a different distribution of heat and cold, moisture and dry weather, among the several seasons. The clearing of lands opens them to the sun, their moisture is exhaled, they are more heated in summer, but more cold in winter near the surface; the temperature becomes unsteady, and the seasons irregular. This is the fact. A smaller degree of cold, if steady, will longer preserve snow and ice, than

a greater degree, under frequent changes. Hence we solve the phenomenon, of more constant ice and snow in the early ages; which I believe to
have been the case. It is not the degree but the steadiness of the cold
which produced this effect. Every forest in America exhibits this phenomenon. We have in the cultivated districts, deep snow today, and none
tomorrow. But the same quantity of snow falling in the woods, lies there
till spring . . . This will explain all of the appearances of the seasons, in ancient and modern times, without resorting to the unphilosophical hypothesis of a general increase of heat. (from Mergen, 1997)

As suggested by Bernard Mergen (1997) in *Snow in America*,
Webster turned a discussion of climate into a political debate—
hard-winter Federalist against warm-trending Republican. Corol-
lary to Jefferson's and Webster's positions on weather are their im-
plicit attitudes toward memory as a source of information past.
Whereas Jefferson trusted the statements of the elderly, Webster was
skeptical and urged better record keeping.

It seems that politics has never been very far from the climate
change debate!

Having considered all the data available to us, it appears that the
warming that is showing up now is too strong relative to recent centuries to be purely a result of natural causes. It is probably that it can best
be explained by something more than natural climate variability. We
have strong evidence that human activity is helping the temperature
component of the Little Ice Age to dissipate, suggesting that a rise in
greenhouse gases in particular is counteracting the general cooling of
the LIA. Research by Michael Mann, from the University of Virginia,
demonstrates very elegantly how temperatures in the Northern Hemisphere have varied in the last few decades relative to the last 600 years.

When the recent rise in temperature seen in the Mann record is
compared with our ice core–generated records of atmospheric circulation, a curious conclusion arises: Atmospheric circulation patterns appear to be within the range of variability of the LIA, but temperatures
over the last few decades are markedly higher than anything during the
LIA (see fig. 5.8).

We are forced to conclude that the LIA is not yet over and therefore
human-induced controls on temperature are at play. While natural climate remains the baseline, human factors may now be overpowering the
trends that natural climate would follow if left undisturbed.

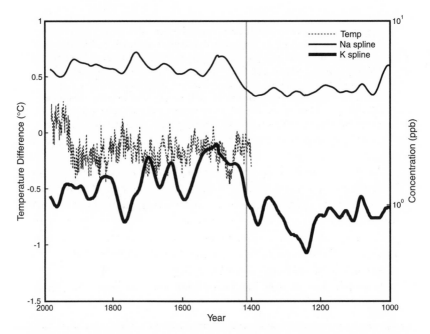

FIGURE 5.8. Natural controls on climate and human controls on climate meet head-on during the twentieth century. This figure compares the GISP2 record of ssNa and nssK (approximations for the strength of the Icelandic Low and Siberian High smoothed using a spline function from Meeker and Mayewski, 2001—see figures 5.5 to 5.7) and change in Northern Hemisphere temperature relative to the turn of the nineteenth century from Mann et al., 1998. Based on the approximation for atmospheric circulation developed from the GISP2 record, the Siberian High and Icelandic Low are still within the range of variability that characterizes the Little Ice Age (LIA). However, temperatures over the last few decades have greatly exceeded those since the inception of the LIA. This may indicate that the recent increase in temperature is not caused by natural controls, but rather by human activity, the new "wildcard" in climate. Superimposing the natural tendency toward LIA atmospheric circulation on a lower atmosphere that is warming, largely in response to increased levels of greenhouse gases, is bound to lead to greater instability in climate.

The implications of this situation go far beyond a simplistic notion of global warming. As previous chapters of this book have shown, natural climate is the product of multiple forcing factors. The modern climate is in turn a product of natural climate plus human-induced variables. The irony is that just as humanity has reached the point where we can begin to track these forcing factors and predict climate change, our own behavior is creating new variables that make it more difficult to know what will happen next.

Another message from our study of the Holocene is, "Don't just think about hot and cold, think about stability and instability of the climate."

It is in this context that we find a reason for environmentally sound behavior that is rarely mentioned: The closer the climate system is to its natural state, the easier it will be to anticipate its future direction. As our human social system throws more variables into workings of the natural climate system, prediction will become an increasingly greater challenge.

Applying this thinking to the Little Ice Age, if natural climate were not being disturbed, we would have a cool troposphere (the lowest level of the atmosphere where most of the gases, molecules, water vapor, and aerosols reside) coupled with generally stormy atmospheric circulation systems. While this combination might challenge humans at times with difficult weather, it would at least retain a certain amount of predictability. Instead, the current situation may be the most complicated in Earth's history to date, because it could be the result of a lower atmosphere (troposphere) whose temperature is to a large degree controlled by human activities (warmed by greenhouse gases and cooled by sulfur emission) combined with significant storminess in the troposphere that is largely controlled by natural factors. Looking back over the climate record, this combination is certainly not natural, and its future evolution is therefore quite complicated to predict.

Multiple Forcing Factors Revisited

As chapter 3 suggests, even the most superficial examination of natural climate confirms that no single factor is responsible for the global system's overall behavior, nor can one cause account for trends in any given direction. Clearly, "multiple forcing factors," as the scientists say, or, in other words, several different climate controls acting together, produce global, regional, and local climatic conditions.

The down side of this realization is the daunting complexity of such a system. Any simple statement about climate is likely to be wrong under these circumstances, and nothing useful can be said about it without a comprehensive inventory of all the forces at work. The up side is that if the nettle of climate system complexity is grasped, some degree (we hope it is a major degree) of predictability will be feasible. If it is not possible to determine a single cause, as a result of all the instability we ourselves are creating, it may be possible to link all the different factors into one description with predictive power on a broad scale.

Juxtaposed against those predictions are trend lines of variables that human civilization produces, such as CO_2 emissions and sulfuric acid. It is in the combination of these two that the promise of useful predictability lies. Regardless of what is done, the climate system will produce what climate scientists call "surprises." These surprises are highly relevant to the everyday existences of human beings living on Earth today.

Catastrophic storms and rapid climate change events are among the surprises that might interfere with modeling and prediction, and more importantly, with the quality of our lives. Still, we can begin to postulate about where the surprises might turn up, which is helpful in itself.

The research conducted to date on the ice cores brought back from Greenland for GISP2, supplemented by observations from other parts of the world, reveals several key factors that interact to affect climate dramatically. Some are completely natural in origin, others completely anthropogenic, and still others a mixture of the two.

Let's take a look at each of these in turn and see how they have varied during the past 1,000 years:

Greenhouse gases. The natural greenhouse effect is, of course, an important mechanism for keeping the Earth habitable. The so-called "greenhouse gases" in the atmosphere allow the sun's radiation to pass through on its way to the surface of the planet, but they absorb much of the energy that the Earth radiates back into space. This process makes the Earth warmer than it would otherwise be, which supports the appearance and continuation of life as we understand it.

Greenhouse gases include water vapor, carbon dioxide (CO_2), methane (CH_4), nitrous oxide (N_2O), and ozone. These gases, along with dust and clouds, are usually found in the troposphere, that layer of the atmosphere closest to the Earth's surface. Naturally produced ozone (so-called "good ozone") is found in the stratosphere. Ground-level ozone (so-called "bad ozone") is found near the surface. It is a human-produced pollutant that can have adverse effects on health.

Greenhouse gases are said to have *sources* (where they come from) and *sinks* (where they go) and they are also transformed in the atmosphere. While some human-caused increases in greenhouse gases might well be absorbed by natural sinks, such as the oceans, we do not yet know enough about the functioning of those mechanisms to predict how well they might play such a role.

The most dramatic evidence of human impact on modern climate is in fact the rise in greenhouse gases in the atmosphere. With the paleo-climate record now available as a basis of comparison, we can say with complete certainty that greenhouse gases have risen more rapidly in the past century than they had at any time in the past several hundred thousand years (and probably much longer) and that this increase is definitely a function of human activity.

CO_2. Carbon dioxide is itself a greenhouse gas that represents only .03 percent of the atmosphere naturally. The ice core record, in combination with the monitoring record, suggests that an increase in carbon dioxide of as much as 30 percent took place from 1850 to 1980, from 265 parts per million per volume (ppmv) to 340 ppmv.

This vital monitoring of modern CO_2 is a remarkable example of how one person can change the course of science. Charles David Keeling of the Scripps Institute of Oceanography started monitoring CO_2 during the IGY in 1957. The monitoring might well have stopped in a few years

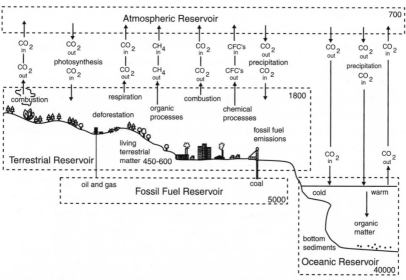

FIGURE 5.9. Sources, transport, and storage of carbon in the Earth system. Carbon is exchanged among a variety of reservoirs that range in decreasing abundance of carbon (measured in billions of tons of carbon) from the oceanic reservoir to the fossil fuel reservoir to the terrestrial reservoir to the atmospheric reservoir.

Deforestation (which decreases the potential for carbon to be stored in plants) and fossil fuel emissions result in 1 to 2 and 5 to 6 billions of tons of carbon, respectively, being added to the atmospheric reservoir each year on average over the last few years. The atmosphere gains approximately 2.5 billions of tons from fossil fuel emissions and deforestation. The remainder is believed to be absorbed by the oceans.

had Keeling not persevered and convinced the government to continue its support of his work on Mauna Loa in the Hawaiian Islands.

As we discussed earlier, by the 1970s and 1980s, monitoring was considered to be unimportant drudgery not worthy of funding, but Keeling knew that we needed a long record to understand the role of CO_2 in the Earth's carbon cycle (see fig. 5.9). Early years of observation revealed a seasonal cycle representing, in effect, "Earth's breathing," plus what was considered to be a minor upward trend. Then, the trend accelerated, and skeptics who originally had thought it to be part of a natural cycle began to back away from their criticisms. Thoughts about the consequences began to trouble scientists and policymakers alike, and these doubts eventually matured into today's global warming debate. If Keeling had given up, we might not have had the valuable early warning that he provided.

≡ RECORD PERFORMANCES

The development of the CO_2 ice core record is no less amazing than the monitoring story. Researchers such as Hans Oeschger and colleagues in Switzerland as well as several French researchers (Jean–Marc Barnola, Claude Lorius, and Dominique Raynaud) demonstrated that greenhouse gases (notably CO_2 and CH_4) were trapped in ice and could be analyzed to produce a long (currently 450,000-year-long) history. Based on their research, we have established a clear link between temperature levels and the relative amount of carbon dioxide in the atmosphere. For example, the French-Soviet ice core record from Vostok, Antarctica, correlated temperature and carbon dioxide very closely, and increased levels of CO_2 are clearly linked with warmer temperatures and vice versa (see figure 2.1).

However, the same record includes sustained levels of CO_2 at the end of the last interglacial period (118,000 to 122,000 years ago) as temperatures decline. This suggests that not every change in temperature is always related to a change in CO_2 during periods of natural variability in CO_2. However, under natural conditions of the last 450,000 years or more, CO_2 levels have not risen as fast or as high as they have in the past century, leaving the question of the relationship between modern levels of CO_2 and modern temperature hotly debated.

What was needed to prove the value of the ice core record of greenhouse gases was a calibration, a meeting point between the values determined from monitoring and those from ice cores. Paramount in this endeavor was the work of David Etheridge from an Australian government research laboratory just outside of Melbourne. He analyzed an ice core from a site where recent greenhouse gas compositions are extremely well preserved. His record from a site known as Law Dome in Antarctica overlapped the monitoring record and showed a perfect transition from trapped gas values in the ice to the monitoring record in the atmosphere. Development of this classic record took years of painstakingly precise work. No average scientist could accomplish such a task and David Etheridge is certainly not that. He bikes 25 miles to work and back each day and is

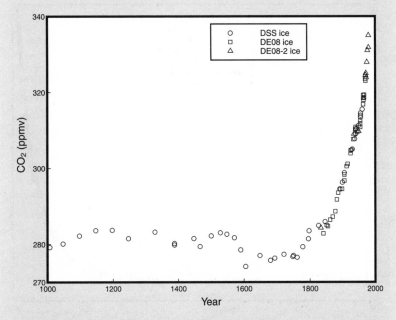

FIGURE 5.10. Carbon dioxide concentrations (parts per million per volume, ppmv) in the atmosphere over the past 1,000 years. Results from ice cores (DSS, DEO8, DEO8–2) in the vicinity of Law Dome in coastal East Antarctica compared with direct measurements of carbon dioxide in the atmosphere demonstrate not only that the ice core measurements provide a true record of change in the atmosphere (the two types of measurements overlap in the mid twentieth century) but also how significantly carbon dioxide has increased in the atmosphere during the twentieth century. *Data from Etheridge et al. (1996).*

a keen competitor in triathlons and endurance races. When I ran with him in Australia, it was a humbling experience. His science is equally awesome.

Based on data from Etheridge, it is clear that the rise in CO_2 has been truly dramatic in the past century, curving upward at an exponential rate. Carbon dioxide levels, however, changed little between the MWP and LIA, further suggesting the involvement of humans in the recent rise. This increase in CO_2 is undeniably paralleled by an increase in global temperatures, as more radiation is retained rather than being beamed back out into space.

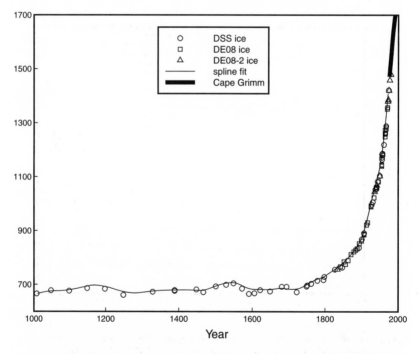

FIGURE 5.11. Methane concentrations (parts per billion per volume, ppbv) in the atmosphere over the past 1,000 years. Results from ice cores (DSS, DE08, DEO8–2) in the vicinity of Law Dome in coastal East Antarctica compared to direct measurements of methane in the atmosphere from Cape Grimm, Australia, demonstrate not only that the ice core measurements provide a true record of change in the atmosphere (the two types of measurements overlap in the mid twentieth century) but also how dramatically methane has increased in the atmosphere during the twentieth century. *Data from Etheridge (1999).*

Methane. As with CO_2, methane is increasing at a rate that has never before been seen, and, as with CO_2, the increase correlates strongly with rising temperatures. The quantities of methane in the Earth's atmosphere have doubled in the past 100 years (see fig. 5.11) from 800 to over 1,600 ppbv (parts per billion per volume). An important cause of methane increase appears to be changes in land use. As temperatures rise, more water vapor leads to the production of more methane, reinforcing the cycle. However, several sources contribute methane to the atmosphere (see fig. 5.12); determining the relative importance and future trend of each is an important task in the understanding of future climate.

Nitrous oxide. Mirroring the behavior of CO_2 and methane, higher levels of nitrous oxide over the last few decades (see fig. 5.13) are also associated with temperature increases. The sources of N_2O in the atmosphere are less well understood than those for CO_2 and methane, but the increase is well documented.

Ozone. As we noted earlier (chapter 2), natural ozone is found in a layer between 15 and 25 miles (25 and 40 kilometers) above the Earth's surface. Ozone functions as a natural shield against ultraviolet radiation, thus exerting controls on a wide spectrum of important matters ranging from climate change to skin cancer.

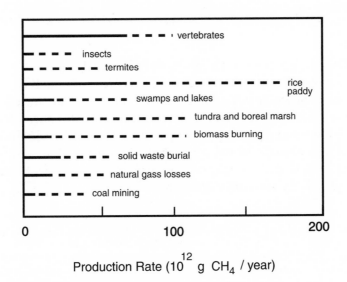

FIGURE 5.12. Sources of methane (CH_4) to the atmosphere. Dashed line indicates range of uncertainty in estimates. *Modified from Khalil and Rasmussen, 1983.*

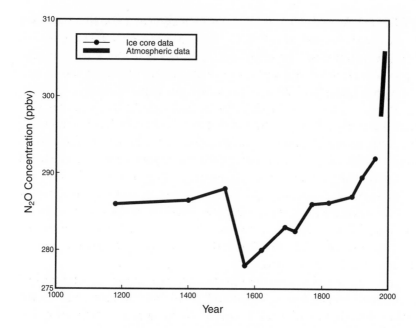

FIGURE 5.13. Nitrous oxide concentrations (parts per billion per volume, ppbv) in the atmosphere over the past 1,000 years. Results combined from several ice cores compared to direct measurements in the atmosphere demonstrate how dramatically nitrous oxide has increased in the atmosphere during the twentieth century. *Ice core data from Pearman et al., 1986; Khalil and Rasmussen, 1988; Zardini et al., 1989.*

Discovery of the springtime ozone hole over Antarctica (as a consequence of the monitoring of ozone that started with the IGY) and the discovery of springtime ozone holes over the Arctic have fueled considerable debate over the cause and consequences of this phenomena (see box). The other type of ozone is a pollutant at ground level, which functions as a reflector of radiant energy, like a greenhouse gas. Current thinking is that ozone may turn out to be a very important variable in the overall climate change equation.

≡ HOW DID OUR UNDERSTANDING OF THE OZONE HOLE EVOLVE?

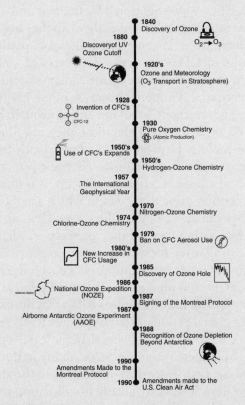

FIGURE 5.14. This figure charts the development of ideas concerning ozone, from its discovery in 1840 through to the policies that have been enacted to curb the destruction of ozone by the 1990s. The scientific knowledge that led to the scientific realization that decreased levels of ozone in the upper atmosphere were harmful to the environment and that humanly produced chorofluorocarbons (CFCs) were particularly effective in destroying ozone required the accumulation of many different scientific discoveries plus monitoring. *Modified from History of the study of atmospheric ozone, Richard S. Stolarski, NASA Goddard Space Flight Center (www.weizmann.ac.il/home/coparthi/ozone/stolarski/history.html).*

Chlorofluorocarbons (CFCs). Unlike some of the other substances that make up the multiple forcing factors for climate change, CFCs are a "no-brainer" when it comes to determining whether they are natural or not. These substances have no natural source, and are purely human in origin (figure 5.14). They were engineered by humans and simply did not exist prior to the 1950s, so we know exactly when they began to have an impact on the environment in general and climate in particular. The chlorine and fluorine in CFCs are active destroyers of stratospheric ozone, which is their primary influence on the climate system.

Water vapor. Water vapor is important because of the key role played by clouds in climate and weather. Clouds can either maintain heat at the surface of the Earth, or at the right elevation, they can function as a shield and produce cooling. Depending on where clouds form, then, they can play different roles in the climate-change equation. Understanding the impact of clouds is extremely difficult. It is impossible to estimate their distribution prior to the era in which satellites became available, though they can be seen by satellites now. They are extremely useful in weather forecasting because they preserve weather patterns when they are present, allowing for weather predictions.

Dust. Like clouds and water vapor, dust plays a dual role, either trapping or reflecting heat. Dust is produced both naturally and as the result of human land-use patterns. Ice core records tell us that dust levels were much higher during glacial periods. Glacial age dust levels recorded at GISP2 are ten to one hundred times greater than those during the Holocene (see fig. 4.2). Still, there is significant variability in dust over the past 1,000 years (see fig. 5.15). Since dust only travels in the atmosphere a few hours to a few days, levels vary greatly from location to location.

Biomass emissions and color. The presence of biomass (organic material) on the Earth makes a region darker in color (decreasing its albedo, or ability to reflect incoming solar radiation). This simple shift in color can affect climate. As an example, when forests replace deserts, the land surface darkens, allowing more solar radiation to be absorbed at the surface. This in turn heats the surface.

In addition, the more plants on the land, the more CO_2 and water is taken up by them, with a corresponding impact on climate. In the ocean, increases in biomass such as phytoplankton lead to elevated emissions of sulfur gases, which leads to the shielding of incoming solar radiation, as described below. Estimates of the biomass over continents and in the ocean for the last millennium are difficult to develop because there is sig-

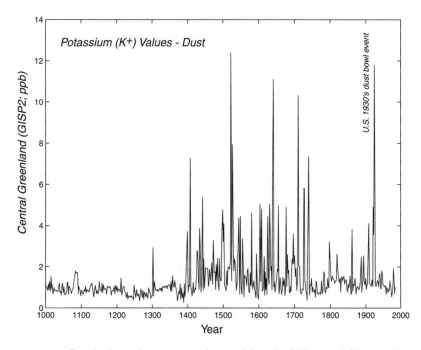

FIGURE 5.15. Dust levels over last 1,000 years developed from the GISP2 record. The 1930s Dust Bowl is clearly seen, in addition to the frequency of such events in the past. Dust levels (background and individual events) have clearly risen since the beginning of the Little Ice Age (A.D. 1400). However, within the context of the last 300 years, the 1930s Dust Bowl is an extremely dramatic event. The increased levels were no doubt a combination of low precipitation exasperated by poor farming practices—a combination of natural climate and human activity.

nificant variability from location to location. We have therefore used several different approaches. For example, GISP2 ice core records of ammonium provide a measure of the average levels of terrestrial biomass in regions upwind from a core site. From these records, it is possible to assess not only the relative amount of biomass over time but also the frequency of forest fires in regions upwind from drill sites (see fig. 5.16).

Sulfur. Volcanoes, the ocean surface (via sea salt and marine organisms), and fossil fuels all act as sources of sulfur dioxide entering the atmosphere. Some of these sources, notably volcanoes, contribute sufficient amounts of sulfur dioxide to form a veil of sulfuric acid (formed when sulfur gases such as sulfur dioxide combine with water vapor) that can shield incoming solar radiation. Volcanic events stand out as prominent spikes in the concentrations of sulfate measured in ice cores. Records of the sulfate deposited in Greenland ice (see figs. 5.17

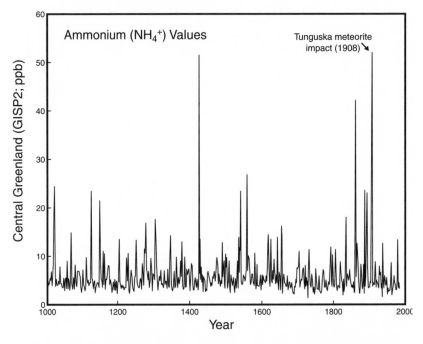

FIGURE 5.16. The past 1,000 years of biomass burning (forest fires) developed from the GISP2 ammonium (NH_4) record. Note in particular the forest fire produced by the 1908 Tunguska meteorite impact in Siberia. Other major periods of forest fire activity are identified by spikes in the record. The GISP2 forest fire record represents a mix of forest fires from throughout the high latitudes of the Northern Hemisphere.

and 5.18) show that concentrations have increased dramatically since the turn of the century because of long-range atmospheric transport of pollution from North America and Eurasia (Mayewski et al., 1986; 1990). The sulfate record is sufficiently detailed to document the decrease in emissions resulting from reductions in industrial activity during the Great Depression of the 1930s and the renewal of industrial activity during World War II. Although the growth of industrial and residential fossil fuel emissions has continued, legislation such as the Clean Air Act of the 1970s has been effective in curbing the release of sulfur from coal burning. This has reduced sulfate levels in the remote atmosphere. Proof of the effectiveness of this legislation is easily found in ice core records.

Earth's orbital cycles. Although the Earth's orbital cycles vary over relatively long time scales, on the order of tens of thousands of years (see figures 3.9 and 3.10), and are best known for providing one of the primary

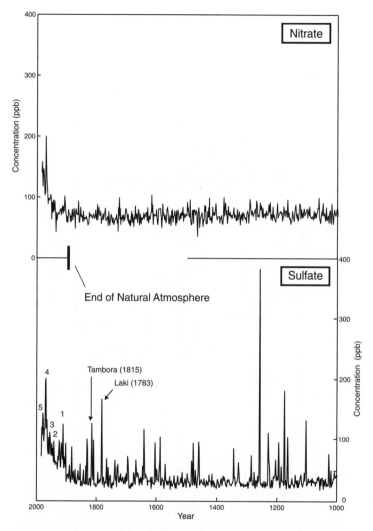

FIGURE 5.17. The past 1,000 years of sulfate and nitrate from the GISP2 record. Note dramatic increase in both of these major components of acid rain during the twentieth century relative to levels of the past 1,000 years. Levels of change in sulfate are closely tied to industrial activity in North America and Europe. On the sulfate graph (1) indicates the beginning of the industrial revolution; (2) the Great Depression; (3) World War II; (4) the period of most intense burning of sulfur-rich, "dirty" coal; and (5) the beginning of the Clean Air Act. The Clean Air Act did not have a dramatic effect on nitrate levels. Most of the short-term (up to one to two year) increases in sulfate are the product of volcanic activity such as the Tambora eruption of 1815 and the Laki eruption of 1783. *Data taken from Mayewski et al., 1986; 1990.*

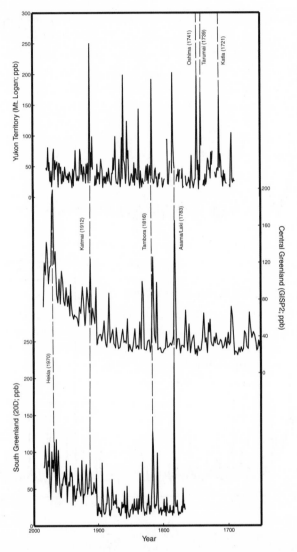

FIGURE 5.18. The past 300 years of sulfate recorded in ice cores from central and south Greenland and the Yukon Territory. Note the twentieth century increase seen in Greenland, downwind from North American and European industrial sources. Mt. Logan is located directly downwind from the Pacific and, although Asia is a large source of sulfate pollution, most is precipitated out over the Pacific, leaving high elevation Pacific coast sites like Mt. Logan with relatively clean air. All three records display volcanically derived sulfate, as noted on the figure. *Data from Mayewski et al. 1993b.*

controls on the pacing of the ice ages, even changes in insolation (energy from the sun) over the last millennium could be sufficient to exert important controls on the length of the growing season and climate change. While the exact magnitude of the change in climate produced by changes in insolation over time and with location is not definitively understood, examination of the last 10,000 years of insolation reveals that, in general, the Northern Hemisphere has received less insolation over this period while the Southern Hemisphere has received more.

Solar output. Scientists know that the sun's energy output varies over time and we believe that these shifts affect climate in some way. However, we need many more observations to comprehend exactly what that effect is. The 11-year sun cycle, for example, has been observed for thousands of years, and such cycles are clearly seen in a variety of paleoclimate records, but the precise sun-climate link still eludes us.

Our current understanding of these cycles is based on observations of the last two solar cycles by satellites orbiting above the Earth's atmosphere. These observations reveal that solar output changes over a solar cycle, but that the changes are relatively weak, accounting for only a .1 percent change in solar energy. While this change in solar output may not be enough to change Earth surface temperatures very dramatically, changes in the chemistry of the atmosphere (notably ozone) also occur during the solar cycle and these could affect climate.

In addition to the 11-year cycle, we have identified several others in the paleoclimate record recovered from ice cores, lake sediments, and tree rings at 22, 78, 208, 510, 1,400, and 2,200 years. The longer the cycle, the harder it is to verify because we need to observe a minimum of three repetitions of a cycle, and satellite observations barely cover three 11-year solar cycles.

Solar variability, long considered controversial as a climate-forcing factor, is now widely accepted to have a role. A recent New York *Times* article stated that, "Today, a growing number of scientists contend that the sun's fickleness might rival human pollution as a factor in climatic change" (Broad, 1997). However, much remains to be learned about the sun's influence before it can be considered the most important factor.

Internal oscillations in the earth system. Does a built-in clock drive the turnover of the oceans (e.g., the Great Salt Conveyor Belt discussed in chapter 3, figure 3.13), and if so, do certain factors change its pacing, thereby affecting climate?

Does the Earth itself have a biological clock that affects climate? In the early 1970s, James Lovelock and Lynn Margulis developed the Gaia hypothesis, named after an ancient Greek goddess of Earth. They used the hypothesis to suggest a mechanism by which the Earth behaves like a living organism. Their primary assumption is that internal controls regulate biological, physical, and chemical interactions to maintain equilibrium in the system as a whole. Aspects of the Gaia hypothesis remain controversial, but it provides some interesting examples of how certain aspects of the Earth system operate, notably the production and maintenance of oxygen and carbon dioxide by living organisms that in turn allow these organisms to survive.

Summary

The climate debate has now come full circle. In the 1960s and 1970s, scientists and policymakers looked at evidence of "global cooling," and worried that the world was entering a new ice age. Today, we are deeply concerned about "global warming," and yet the evidence of the Ice Chronicles tells us that we may still be in the Little Ice Age!

The discussion continues as to whether the modern controls on climate are primarily natural or human in origin, but a careful analysis of the multiple forcing factors known to affect the climate system convincingly argues for a combination of the two. In any event, we know that an either/or debate of this kind is not very productive. The future of the climate system will be determined by the interaction of both natural and human-produced factors, and we have no choice but to comprehend the changes ahead, learn to predict them as best we can, and try to adapt where we cannot do otherwise. We can begin this process by understanding the impact on people and society of the kinds of instability and uncertainty that are being produced in the climate system today.

6

Climate Change: The Real Impact

Global warming is already beginning to take a toll in Alaska, on the forests, in the loss of salmon habitat, and in wide-ranging melting of permafrost that is damaging roads, houses, and airports, scientists say. After years of debate over the reality and extent of global warming, says Glenn Juday of the University of Alaska, "It's not just projections anymore. It's an unfolding reality . . . If the damage to forests continues as severely as it has been, there is a serious question of whether any kind of forest resembling what we've got now will continue," he warned. Juday, a professor of forest sciences, said the effects of global climate change on the Alaska environment in the last 20 years are becoming evident. That's a period when the state—along with many other parts of the world—experienced a "regime shift" in which conditions began to change noticeably.
—David S. Chandler, *Boston Globe*, 1997

Ultimately, climate change is important to all of us, not because of something that happened a million years ago, a thousand years ago, or even a hundred years ago. Great fascination with paleo-climate studies, while the information is available to anyone with curiosity, is generally limited to the scientific community. The majority of people are concerned about climate change because of its impact on weather patterns on a daily, weekly, monthly, or annual basis.

Because people are interested in weather, extreme weather events such as major storms confirm the growing belief by non-scientists that weather really does change in response to climate change, regardless of how the debates among scientists and policymakers turn out.

✳ THE DANGER OF DANCING SNOW

As I looked across the ice sheet, I saw something that put all my senses on the alert. It was during the 1971–1972 field season, my third in Antarctica, and the last for my Ph.D. research. I was exploring the mountainous region of Southern Victoria Land with a colleague named Bob Wilkinson when a storm hit. (I've lost touch with Bob, but the last I heard, he was working as an Alaskan bush pilot.) At the time, we were trying to get from one mountain to another, hauling our sleds across the ice.

A faint swirling of snow on the ice sheet heralded an approaching storm. Even though I was still a relative newcomer in the field, I had quickly learned that what I called "dancing snow" meant trouble, because it signaled that the atmospheric pressure was dropping, and a major storm would not be far behind. Not long after I saw that little snow-dance, the wind did pick up, and it soon gusted to 100 to 120 miles per hour (160 to 190 kilometers per hour).

Sometimes, Antarctica gives you options, sometimes it does not. In this case, we had no choice about how to respond. We stopped right there, pitched our tents, and prepared to wait the storm out. I couldn't believe the strength of the winds—we had brought along a 120-pound generator that kept jumping around on its platform, and we needed every piece of climbing rope and all of our extra climbing screws to tie down the generator and our tent.

We knew all too well that storms like this one could last up to ten days, so we couldn't predict how long we would be holed up in our tent. As it turned out, we were trapped for three days, completely out of contact with our colleagues at the main base, McMurdo Station, 100 miles (161 kilometers) away. Rescue was unthinkable; we couldn't expect the people there to risk their own lives coming out in the storm to find us. We took turns holding up a piece of plywood against the wind wall of the tent to keep the whole thing from collapsing. That fabric was all that stood between us and a very serious situation. Without shelter, we would have been subjected to temperatures that hovered around o degrees Fahrenheit (−18 degrees Celsius) and that was without the wind chill factor!

We really couldn't afford to risk going to sleep, because we had to be sure the tent remained intact, so we were awake for most of those seventy-two long hours. What do you think about in a situation like that? I just focused on how long the tent would stay up and (therefore) how long I would stay out of peril!

On reflection, I guess I also found myself hoping we would learn something that would help other people, since that's ultimately why we were there.

Finally, the storm cleared, and we were flown by helicopter 100 miles to the main U.S. scientific station in Antarctica, McMurdo Station. We had experienced firsthand one of nature's potential catastrophes.

For people living in coastal areas, the more extreme predictions about global warming are truly frightening. For them, the idea that warming trends might cause rising tides that will wash away their houses has immediate, practical implications. The complicated arguments about what is causing warming or cooling to occur are less relevant than the practicality of their situation.

These people really just want to know, "What does it all mean to me?" Island inhabitants are even more vulnerable than those living along continental coasts. Residents of such countries do have the option of moving inland. However, many nations are basically islands, and if these islands are washed away by rising sea levels, their citizens will become stateless refugees. "Eco-refugees" might turn out to be a serious political and economic problem in the future. We know from the American "dustbowl" experience of the 1930s that "eco-migration" is a reality when extreme weather hits entire regions for long periods of time.

Similarly, the farmer in the midwestern United States or the Horn of Africa depends on predictable weather patterns to make a living. Within very recent memory, such areas have been tortured by long droughts so severe that precious topsoil has withered and blown away. Meanwhile, hurricanes, tornadoes, and other severe weather phenomena are highly destructive to human life and property around the world. In July 2000, NBC news correspondent Robert Hager reported that

"It's been a summer of extremes with wild weather around the world and the forecast of many more changes in years to come" (Hager, MSNBC, July 25, 2000). Hager pointed to 100-degree-Fahrenheit (37-degree-Celsius) temperatures in the American South and Southwest, while heat of 120 degrees Fahrenheit (49 degrees Celsius) triggered roaring brush fires in Greece.

Hager coupled these reports with a story about a decade-long NASA study that showed Greenland's ice sheet thinning at the rate of 1 yard (.91 meters) per year at the water's edge, a total loss of 50 billion tons of water.

While our technological prowess has helped us to predict severe weather events like these, and therefore reduce the loss of life, we have made little or no progress in preventing them or altering their behavior. Hager's report also made an important point regarding policymaking and climate change. He noted that, while the United States had experienced extreme weather over the summer, it had been much worse worldwide. This is important because the United States exerts so much influence in policymaking on climate change, but may be less aggressive than other countries because it hasn't fully experienced the impact of devastating weather as yet.

≡ ICE STORMS AND HURRICANES: DRAMATIC DAMAGE AT LOCAL TO REGIONAL SCALES

When we think of disastrous weather, ice storms are not typically at the top of the list. However, they can be deadly. For example, the storm that lasted from January 15 to January 18, 1998, devastated parts of New England, New York, and Quebec. It was truly a major disaster, even though it was not the worst ever to hit the region. In November 1921, for example, freezing rain fell over the same region for 75 hours.

However, despite the greater natural force of storms prior to the 1998 event, this particular event was the most destructive on record simply because there was so much more to destroy. Population has increased in the region, as has technological infrastructure. One of the main problems was that major power lines thought capable of sustaining the weight of ice loading buckled, plunging thousands into darkness and cold for days on end.

Devastation to the ecosystem was also immense. Trees snapped like toothpicks over vast areas; they will probably remain as visible reminders for decades as evidence of the storm's incredible force. Map reconstructions of the forest damage reveal that the destruction increased with elevation and distance inland from the ocean— colder temperatures in these areas resulted in increased ice buildup.

Hurricanes are a different kind of disaster, and one that we have grown to fear as we have learned more about them. Typically, these storms have a diameter of close to 400 miles (644 kilometers) and maximum sustained surface winds of 74 mph (119 km per hour). They may extend to 40,000 feet (12,000 meters) upward in the atmosphere. The largest hurricanes (or typhoons) are found in the Pacific, but most of the tropics and mid-latitudes also experience the wrath of hurricanes—most typically in late summer to autumn. In spite of tremendous advances in prediction and early warning capabilities, the eighty or so hurricanes that form each year result in, on average, 20,000 fatalities and immense property damage (Barry and Chorley, 1992).

It is clear that if climate change were to result in increased frequency of hurricanes, we would see much more death and destruction as a result.

Hurricane frequency varies from region to region. In New England, for example, the storms tend to strike more often in the south than the north because they build their strength over the relatively warm waters of the Gulf Stream as it deflects eastward near Cape Cod. However, the most intense period of hurricane activity over New England appears to have been the late nineteenth century, with a decline in the number hitting the region in recent years (Easterling et al., 2000).

In the introduction to this book, we considered the shift in spatial consciousness that has resulted from human beings venturing into outer space, and having the opportunity to look back at the Earth from that new vantage point. We also noted that studying the Ice Chronicles triggers a complementary change in perception of time. Looking at even 110,000 years of climate history reminds us of how very old the Earth really is, and that it has experienced many, many shifts in climate during its lifetime.

To the astronauts, the Earth looks fragile when seen from outer space, and fragility was the term most often used by them to describe the planet when viewed against the backdrop of an infinite and eternal universe. However, the Earth itself is far less delicate than many imagine. Having existed for billions of years, it seems designed to last for billions more—barring complete and utter destruction by a cosmic catastrophe, such as collision with an enormous meteorite.

Climate change in itself does not threaten the existence of the Earth, though it may be dangerous to current urban human systems now existing on the planet. It may very well be that the part of the ecosystem supporting human life is properly viewed as fragile and vulnerable. And the climate, which is intimately connected to this ecosystem, is easily and quickly disturbed, as the preceding chapters have demonstrated.

Nevertheless, our studies of climate and civilization over the millennia offer us great hope because of all that we have learned about the adaptability of human beings, both now and in the past. Lessons from earlier times suggest that we still have choices about the impact of climate change. In some instances when rapid climate change events have struck, it appears that civilizations may have been destroyed by unforeseen changes beyond their ability to respond. In other cases, people adapted effectively to the new conditions and flourished.

For example, it seems that the Norse in Greenland stubbornly clung to their European way of living when the Medieval Warm Period suddenly transformed itself into the Little Ice Age. The Greenland colony dropped from history, while the Inuits may have only made slight shifts in their behavior to adapt to the new circumstances, and survived.

Similarly, we mourn the disappearance of the urbanized Mayans during the Middle Ages, and wonder what made them abandon their great cities in Central America. But Mayans still live on the same ground, and they still have a proud culture. Is it possible that their ancestors simply adapted to rapid climate change, and did so in an effective way that we don't understand?

Considering this broad perspective of the Earth's history and the human place in it, and stepping back from the details revealed by the Ice Chronicles about forcing factors of climate change, what is the big picture that emerges? What do these records reveal about the real impact of climate change on human beings and their societies?

The first, and probably most important, insight is that if we assume we will somehow find a technological way to simply fix what we consider to be broken in the global environment, that will turn out to be a very dangerous assumption. The climatological system is complex, and we are only just beginning to understand it at a time when our own behavior is making everything more difficult to sort out. Believing that there is an easy "fix" allows us to continue our current behavior, acting more like the Norse and less like the Inuits.

The second insight is that human impact on the environment has varied significantly over time. Initially, our activities affected the landscape, which did not produce a drastic response in terms of climate change. However, with the advent of advanced industrial civilization, we have begun to affect more of the environment. The result is that we are having a profound impact on the climate system. That much is certain from a study of the Ice Chronicles.

Human beings can live in harmony with the natural environment, including the climate system. Industrial society, which is a relatively recent phenomenon, has clearly had a tremendous impact on natural climate and the chemistry of the atmosphere, by increasing the greenhouse effect, among other things. However, we may well be moving into a post-industrial period, in which the "smokestack" economies of the past are no longer dominant. If this is so, it may turn out to be a modern example of positive adaptation to climate change, even though that was not the original reason for developing these new technologies.

Right now, the issue is one of immediate importance because many less-developed countries are tempted to follow the Western path to prosperity through industrialization. If another way can be found to provide them with high standards of living without the pollution that accompanied the industrial development of the nineteenth and twentieth centuries, that will be good for those countries and the world in general. For example, Harvard University's Interfaculty Initiative on the Environment includes a "China Project" addressing precisely this issue. In this way, the prospect of global climate change has the unintended consequence of accelerating global cooperation. Regardless of political differences, it is in everyone's interest to find a solution to this problem that is acceptable to developing and developed nations alike. Reducing the impact of global warming could become a unifying project for our emerging global culture (McElroy, 1997).

The third insight is that human beings must continue to adapt to climate change, since change is inherent in the system, as the 110,000-year Greenland record clearly demonstrates. Rapid climate change events

(RCCEs) do not depend on human-produced pollutants to be triggered and to transform local, regional, or even global climate in the space of a few years.

Humans cannot yet affect the natural tendency of the climate system to produce RCCEs, but we do have a choice about how our economic and social systems will accelerate, retard, or eventually alter those developments. The rise in greenhouse gases illustrates a clear-cut instance in which the human system is a variable in the natural climate system. However, as chapter 5 demonstrates, this well-known example is only one of several ways in which the human influence on the total environment is spreading worldwide. As just one example, with more people moving to the cities, they contribute to the "urban heat island effect." Cities are warmer than rural areas, trapping heat that would normally be radiated back into space, much as greenhouse gases do.

≡ FROM URBAN HEAT ISLANDS TO NUCLEAR WINTER

Scientists have demonstrated that urban centers have a major impact on local climate in a variety of ways. Cities add more relief to the landscape, usually resulting in a decrease in wind speed. Cities are also typically sources of moisture as a consequence of industrial activity and general energy consumption. Urban emissions of sulfate aerosols from fossil fuel burning and dust production caused by industrial activity and human occupation result in poorer visibility and increased frequency of fogs. Fortunately, decreases in emissions over the past two to three decades, resulting from clean air legislation, have resulted in a decline in urban fogs and more sunshine in some cities, but increased population and industry near urban regions could jeopardize that trend. Countries that do not legislate clean air continue to have poorer and poorer air quality each year.

Concentrated energy consumption in cities leads to warming relative to rural areas. Major cities such as Chicago, Washington, Los Angeles, Paris, Moscow, Philadelphia, Berlin, New York, and London record temperatures that range from 1.08 to 2.34 degrees Fahrenheit (.6 to 1.3 degrees Celsius) higher than surrounding areas (Detwyler, 1971).

At the other end of the spectrum, we can contemplate some human-induced climate changes to which adaptation would be difficult, if not impossible. Perhaps the most dramatic suggested alteration to climate as a result of potential human action is the nuclear winter hypothesis. First proposed by the Nobel Prize–winner Paul Crutzen and his colleague J. W. Birks in 1982, it paints a frightening picture of the climatic consequences resulting from the large amounts of dust and smoke that would be produced by the fires that would follow nuclear detonation (Crutzen and Birks, 1981).

A series of national and international committees met to assess the consequences of large-scale nuclear warfare in the context of the hypothesis. Although based solely on models, the theory suggests that the resultant shielding of incoming solar radiation could trigger a cooling that would last years or even centuries. Analogs can be found in nature for such an event, notably meteorite impacts and major volcanic eruptions. For the latter, the Ice Chronicles offer a unique perspective.

About 71,000 years ago, the island of Toba in Sumatra literally blew up, ejecting about 1,000 to 1,200 square miles (2,500 to 3,000 square kilometers) of rock into the atmosphere. Sulfate produced during this eruption is found in the GISP2 ice core. The eruption was almost 100 times greater than the largest known historical eruption (Tambora, A.D. 1815) that produced "the year without a summer" in eastern North America and western Europe. Toba was so enormous that it probably left sulfate aerosols in the atmosphere for a minimum of five to ten years and perhaps significantly longer, as documented by Gregory Zielinski of the University of Maine (Zielinski et al., 1996). As a consequence, it could have led to ocean surface cooling and climate cooling observed in the GISP2 ice core record that lasted several decades. If a volcano could have that effect, imagine the impact of multiple nuclear weapons.

We can also examine the effect of climate change on society by considering recently published reports. Documents of this type are a relatively recent innovation in science for a variety of reasons. First, widespread interest in climate change has itself been recent, largely as a consequence of media coverage surrounding not just the "global warming" debate but perhaps even more importantly the direct consequences of climate change, that is, loss of life and property. Second, it is only in the past several years that data covering sufficient time periods and over large enough geographic areas has been available for study.

An Example of a Global View of Climate Variability

Researchers from the Climatic Research Unit at the University of East Anglia in the United Kingdom have been leaders in analyzing climate records. Their most recent synthesis reveals how annual average temperature over land and ocean has changed since the earliest period of widespread instrumental records in the mid-nineteenth century.

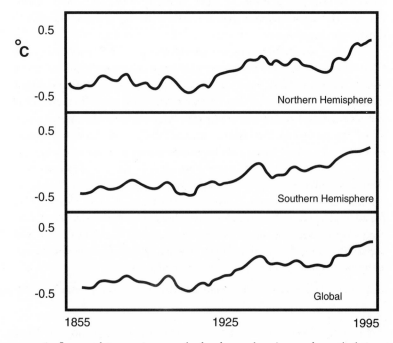

FIGURE 6.1. In general, temperatures over land and ocean have increased over the last century. This figure shows temperature records since 1850 for the Northern Hemisphere, Southern Hemisphere, and global. The data set is a combination of land air temperature data (Jones, 1994) and sea surface temperature data (Parker et al., 1995). The data are expressed as anomalies relative to the period 1961–1990. *Figure modified from the University of East Anglia Climatic Research Unit website: www.cru.uea.ac.uk/cru/data/temperat.html.*

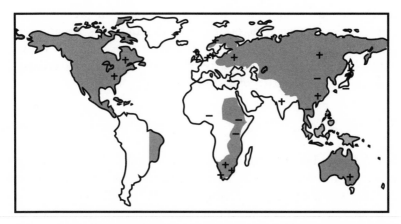

FIGURE 6.2. Regions for which daily precipitation data are available (Easterling et al., 2000). For most countries, these data have been available only since World War II. For Australia, United States, Norway, and South Africa, the data began near the early twentieth century. Positive and negative signs indicate regions with particularly heavy or light precipitation over the past few decades. Only shaded areas were included in this survey. *Figure modified from Easterling et al., 2000, p. 418, figure 1.*

Viewed from the perspective of the Northern and Southern Hemispheres and globally, the overall increase in temperature of the last century is clear, as are departures from this trend such as the cooling during the period from about 1940 to 1970 (discussed in chapter 5).

Typically, it is the extreme climate events that demand our attention. Recent reports from researchers such as Tom Karl and David Easterling at the National Oceanographic and Atmospheric Administration (NOAA) and their colleagues at other institutions reveal the variability and trends in extreme climate events. As Easterling pointed out in a recent article, the flooding associated with Hurricane Mitch in 1998 left more than 10,000 dead in Central America, and mainland U.S. hurricanes resulted in $5 billion in damages in 1995—devastating eye-openers (Easterling et al., 2000). Their research pinpoints the regions of the world that appear to have experienced an increased or decreased frequency of extreme events.

As these researchers point out, data is not available for the entire world, and even when it is available, few regions have detailed coverage prior to World War II. Still, it is possible to draw some very interesting conclusions from their research. Their analysis provides important information, for example, concerning precipitation extremes. As demonstrated in figure 6.2, significant changes in precipitation extremes exist in several parts of the world. Increases are apparent in North America, Europe, northern Asia, Australia, and southern Africa, while decreases are revealed in portions of southern England, central and east Africa, and parts of eastern Asia.

Land Use and Climate Change

It isn't just industrial activities that affect climate. We also change the albedo, or reflectivity of the Earth's surface, by clearing forests or shifting land uses in other ways. Dating back several millennia, humans have been altering their environment through such land use changes. Now, as billions of human beings crowd the planet, the attendant requirements for food and raw materials mean that land use alteration is greater than ever before. As summarized by William R. Cotton and Roger A. Pielke in their book *Human Impacts on Weather and Climate Change*, the two most important impacts on climate produced by changes to the landscape are those related to transformations in albedo and in moisture available to the atmosphere through evaporation and transpiration (Cotton and Pielke, 1995).

In general, we produce major albedo shifts as a byproduct of deforestation (causing a decrease in the exposure of dark surface area), cultivation of crops (causing a change in surface color because of crop color and original surface color), construction of large paved surfaces (in the case of airports, for example, leading to a lighter-colored surface), and desertification through poor agricultural practices (resulting in an increase in light-colored surface) (Karl and Easterling, 1999).

We produce changes in moisture balance as a byproduct of desertification (decrease in moisture), irrigation (increase in moisture available for evaporation), and creation of human-made lakes (increase in moisture available for evaporation) (Karl and Easterling, 1999).

An Example of a Regional Scale Prediction for the United States

The challenge faced by climate researchers in filtering out natural from anthropogenic causes of climate change is to create even finer analyses of climate change in both time and space—in time by bringing more of the analysis down to decade-by-decade or year-by-year, and in space by regionalizing the data. Policymakers can then consider their decisions with an equal attention to detail, rather than using broad strokes that may be too sweeping to attack the real problems.

Several different types of records are available for constructing these yearly regional views of climate change. Instrumental records are the best because they provide a direct measure of temperature, precipitation, wind speed, wind direction, humidity, cloud cover, and so on. However, paleoclimate records are now being used to provide information for periods preceding the instrumental era of climate observation. Tree-ring studies, for example, chart the periods when regions have experienced variations in temperature and droughts.

Historical records, tropical corals, and ice cores provide El Niño Southern Oscillation (ENSO) information. Historical records from South America document drought and flood associated with phases of ENSO. Corals reveal information about winds and precipitation associated with ENSO. Tree rings and ice cores also contribute to understanding atmospheric circulation, while ice cores tell us much about changes in sea ice.

The U.S. government has moved in the direction of regional climate prediction with a report called *Climate Change and America*, which forecasts trends over the next century for seven regions of the country. The report may or may not turn out to be accurate, but it is fascinating in revealing the kind of impact that may occur if some of the current trends remain unchecked.

A summary of the early drafts of the report reveals the following predictions for each of the regions:

Northeast. This region is predicted to experience warming from 4.5 to 9 degrees Fahrenheit (2.5 to 5 degrees Celsius) by 2100. The major resulting problems could include rising sea levels that might, for example, flood New York's subway system, and put JFK airport out of commission. The report suggested that New England will eventually experience many more ice storms, and that the ski industry may not remain viable.

Southeast. This region may experience destruction of coastal trees by the incursion of salt water, and flooding could destroy many levees, bringing devastation to riverside communities. Droughts could reduce some crops, such as corn and rice, but more timber might also be grown under the warmer conditions.

Midwest. Predicted to have the greatest amount of warming, the Midwest may experience a lengthening of the growing season in the north, and a drying out of soils. The urban heat effect could make summers less healthy in larger cities, while lake levels could also decline.

Great Plains. The prediction of more droughts in this area would mean fewer family farmers, as only the larger, stronger farms would survive the hard times. Increased heat in this region could disrupt the flight of migratory birds, with unknown consequences.

West. Heavier winter rains could increase the already dangerous possibilities of mudslides in certain areas. The snow pack in the mountain ranges might melt, creating numerous problems. Fruit and nut crops could decrease because they rely on winter's chill for growth.

Pacific Northwest. Melting mountain snow packs could create urban water shortages. Tree-killing pests and forest fires may increase with the predicted warmer and drier summer months. More flooding is forecast for the winter months.

Alaska. The landscape could be transformed as winters become 18 degrees Fahrenheit (10 degrees Celsius), warmer, melting the tundra and causing caribou and reindeer to starve. The fish population could fluctuate wildly, but oil and gas exploration may become easier because of milder weather.

The first draft of the report went out to some 400 reviewers, many of whom found it too negative. Some noted that there would clearly be "winners" and "losers" as different regions responded to the predicted alterations in weather patterns. Later versions of the report included more references to the possibility that the predicted changes might benefit some while hurting others. (Fialka, *Boston Globe*, 2000)

☰ **FROM GREENHOUSE GAS WARMING TO COOLING IN THE NORTH ATLANTIC: SURPRISES IN THE CLIMATE SYSTEM**

As we have observed from the GISP2 ice core record, major rapid climate change events (RCCEs) have occurred throughout the last 110,000 years, and probably throughout Earth's climate history. Comparison of deep-sea sediment records with the GISP2 RCCEs tells us that many, if not all of these events, were associated with a shutdown of the world ocean's "Great Salt Conveyor Belt" (see figure 3.13). Each shutdown of the conveyor led to cooling over the North

Atlantic, which is revealed through the GISP2 stable isotope estimate for temperature.

A variety of possibilities may explain the shutdown, ranging from changes in the extent of sea ice in the North Atlantic that could alter the heat balance of the surface ocean, to changes in the salinity of the North Atlantic. In a greenhouse gas–warmed atmosphere, both sea ice extent and ocean salinity could vary dramatically.

Climate modeling by Syukuro Manabe and Ronald Stouffer of the Geophysical Fluid Dynamics Laboratory, NOAA, at Princeton University has revealed a potential surprise association between greenhouse gas warming and the conveyor belt (Manabe and Stouffer, 1995). Their model's calculations suggest that an increase in fresh water to the North Atlantic produced by increased precipitation in this region, coupled with melting of snow and ice (as a consequence of greenhouse gas warming), could contribute sufficient fresh water to the North Atlantic to alter its salinity. This might be sufficient to slow down or shut down the conveyor in this region. In so doing, heat transport across the North Atlantic that is driven by the action of the conveyor would decrease dramatically. The potential surprise, then, is *cooling* of the regions downstream from the conveyor, such as northern Europe, as an initial consequence of greenhouse gas warming.

≡ THE IMPACT OF INSTABILITY IN THE ATMOSPHERE AND OCEAN: EL NIÑO AND THE SOUTHERN OSCILLATION (ENSO)

While the terms *Holocene, Medieval Warm Period,* and *Little Ice Age* are still alien concepts to most people, the terms *El Niño* and *La Niña* are suddenly household names. This is because instabilities in the natural climate system and potential instabilities created by human controls on climate may be having an effect on these natural weather patterns, to the detriment of human beings worldwide.

The El Niño Southern Oscillation (ENSO) is a combined ocean-atmosphere phenomenon that exerts a strong influence on global climate. This system is now notorious for bringing to bear a multiplicity of negative effects on human society. Typically a Pacific

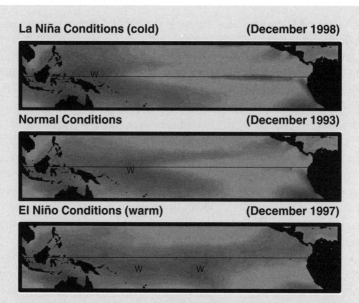

FIGURE 6.3. The El Niño and La Niña phases of the combined ocean and atmosphere phenomena known as the El Niño Southern Oscillation (ENSO). Equatorial Pacific sea surface temperatures during a normal period (e.g., December 1993) are characterized by cool water in the Eastern Pacific and warm water in the Western Pacific. During a strong La Niña event (e.g., December 1998) the Eastern Pacific is cooler than usual and cool water is observed farther west than during normal conditions. Strong El Niño conditions (e.g., December 1997) are characterized by a broad band of warm water extending across the equatorial Pacific. Dark-shaded regions with W denote warmest water. *Figure modified from www.pmel.noaa.gov/toga.*

Ocean phenomenon, El Niño occurs with some regularity (every 3 to 10 years). ENSO can bring too much rain to some regions, severe drought to others. Ocean temperatures in the tropical Pacific can rise to the point where dramatic changes in fisheries occur as well, and it was fishermen who first gave the phenomenon its name (El Niño means "the child," and refers to the Christ child because fishermen from South America initially noticed it occurring in December). Interestingly, climate records of the past 1,000 years show that the effect appears more frequently during times of natural climate cooling. For example, it appears that fewer El Niño events occurred during the Medieval Warm Period than during the colder times that preceded and followed it (see fig. 6.4).

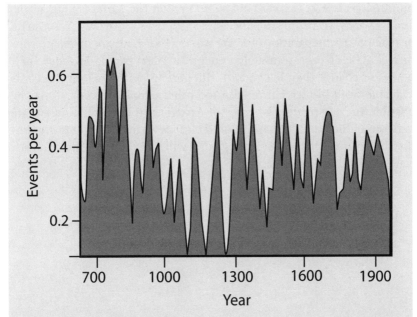

FIGURE 6.4. Frequency of El Niño events over the last 1,400 years, modified from Anderson (1992). According to this reconstruction, more El Niño events occurred during the Little Ice Age (since A.D. 1400), when temperatures throughout much of the world were cooler, than during at least the latter part of the Medieval Warm Period (A.D. 1,000 to A.D. 1300), when temperatures were generally warmer. However, El Niño events seem to have increased during the twentieth century as temperatures have risen. How much has human activity affected El Niño? Is it solely a naturally occurring phenomena? *Figure modified from Anderson, 1992, p. 194.*

In the 1990s, however, El Niños have been appearing annually and lasting longer, with significant effects on regional climates. It remains unclear whether human activities have affected the ENSO system and whether increasing temperatures, rather than decreasing temperatures, now drive the frequency and strength of El Niño events.

Summary

People around the world care about global warming and climate change because these are more than scientific fields of inquiry—they are matters of life and death for many of the planet's citizens and societies. One hundred years ago, we were unable to predict the path of a hurricane, and hundreds of people died because they had no warning to evacuate or make other preparations for the storm. Today, we are unable to predict climate change accurately enough to warn people to move their homes or change their livelihoods. But perhaps with the longer records available from the Ice Chronicles and other climate records, it may be possible to investigate past cases of extremes in climate, differentiate between natural and human-induced climate change, and improve predictability. If we can achieve this goal, it will have positive effects on society that will be immense.

7

Confronting the Choices: Scientists, Politicians, and Public Policy

In an earlier chapter, we referred to the impact of politics on the embryonic science of climatology during the formative days of the American republic. Two hundred years later, the tools and techniques of science have become far more sophisticated, while politics have become global and far more complex. However, the interaction between scientists and politicians to create public policy continues unabated, and much is similar today to the dynamics of those earlier times.

The Ice Chronicles give scientists a great deal to think about, and are also a powerful policy-making tool. However, as we ponder what these records are telling us, and what should be done about their message, we must bear in mind that different political views are always clashing internationally and domestically over climate change in general and global warming in particular. These political perspectives cannot be eliminated in determining the best climate change policies. Indeed, they must be incorporated into the debate.

SEEING CLEARLY

We usually travel into remote sites in Antarctica on board Hercules LC-130 aircraft. The National Science Foundation owns these planes, and up until recently they were operated by the U.S. Navy. Now, they are operated by the 109th Air National Guard from Scotia, New York. The 130s are remarkable. For "open field" landings (where there is no prepared surface), they can fly in and drop off up to about 20,000 pounds of personnel and equipment. That's usually more than enough to equip our 100-plus-day traverses with everything we need:

FIGURE 7.1. C-130 aircraft landing with skis at a tent camp in the Transantarctic Mountains, Antarctica. *Photo by Paul Andrew Mayewski (1982).*

FIGURE 7.2. First U.S. traverse of Northern Victoria Land, Antarctica. The snowmobile team is headed toward a small mountain range to study glacier activity as an indicator of climate change. Small glaciers like the one in the center of the picture are more sensitive to change in climate than the inland portions of the ice sheet onto which they flow. The small glacier is approximately .62 miles (1 kilometer) across and up to 656 feet (200 meters) thick, while the ice sheet on which the snowmobiles are pictured is approximately 1 mile (1,500 meters) thick or greater. This small mountain glacier and many others in the region have experienced retreat over the last few decades due to milder climate conditions (Mayewski and Attig, 1978; Mayewski et al., 1979). *Photo by Paul Andrew Mayewski (1975).*

FIGURE 7.3. Ice core boxes being loaded onto the rear of a C-130 aircraft during glaciological reconnaissance activities in West Antarctica. *Photo by Paul Andrew Mayewski (1992).*

snowmobiles, food, fuel, scientific equipment, and camping gear. They are the only such aircraft in the world equipped with landing skis, which gives the U.S. Antarctic Program (USAP) an amazing degree of access to most of the vast continent of Antarctica and the ability to bring in and out of the field large amounts of equipment and samples.

When the aircraft lands in one of these remote sites, typically hundreds of miles from the main base, it is cold, at a high altitude, and above all, extremely noisy. The noise and blowing snow are produced by the four aircraft engines, which must be left running the entire time because it is so cold—shut them down, and they may not start again. The personnel and equipment are off-loaded from the tail of the aircraft.

Then, the exhaust fumes, noise, and flurry of activity are suddenly gone as the aircraft departs for home base at McMurdo Station. We stand and watch the plane with mixed feelings, as our last physical contact with the outside world seems to dwindle in size and disappear.

Standing there, the first thing you notice is the quiet—it doesn't seem possible to have so little sound. After all, there isn't even the proverbial "tree falling in the forest" that one might or might not hear. You look out over the vast expanse of ice, which at first seems featureless. Then, you notice that the landscape does indeed have distinct features. For example, there are small ridges and valleys in

FIGURE 7.4. Expedition members preparing snowmobiles to be removed from the rear of a C-130 aircraft as part of the glaciological reconnaissance activities in west Antarctica. *Photo by Paul Andrew Mayewski (1995).*

the surface snow or carved-out hollows from the wind. At some of our landing sites, we can observe mountains far off in the distance. The realization that the mountains are sharply defined, despite being tens of miles away or even farther, is striking. You understand that the view is sharper, more distinct, than any such vista back home. In fact, the view is so clear that from the right vantage point, you can see the curve of the Earth, one degree of latitude distance (62.5 miles, or 100.6 kilometers).

Why can you see so clearly in Antarctica? Because this is the most pristine atmosphere on Earth. There is barely even any dust from natural sources like deserts. At first, you imagine that such a clear view is unique to Antarctica. Is this what the atmosphere should look like? Yes, it is, and without the pollution of our modern society, much of the planet would offer clear views like this one. The juxtaposition of the Antarctic perspective and the first sighting of our home continent, as we land in Los Angeles, is enough to convince anyone that we have lost something precious at the expense of modern society. To be sure, we couldn't land at Los Angeles International Airport without creating some amount of pollution, but these expeditions do make you wonder—how much should we compromise?

Science tries to do its part in the climate change debate by providing good information to politicians and policymakers. For example, since its founding in 1988, the Intergovernmental Panel on Climate Change (IPCC) has taken on the role of assessing the scientific and technical merit of information pertaining to climate change in an objective, comprehensive, and transparent manner. To accomplish its mission, IPCC organizes groups of scientists and technical experts representing a broad spectrum of nations and expertise. The expertise comes from academia, private organizations, national research laboratories, and nongovernmental organizations (NGOs). Thousands of experts are called upon to prepare, review, and edit a report that describes the current state of knowledge with respect to climate change. The IPCC's *Second Assessment* came out in 1995 as a three-volume compendium. The assessment is extremely important because it represents a consensus view—perhaps as objective a view as is possible to achieve within the global scientific community. If IPCC makes a statement, it can be considered to be acceptable to the vast majority of scientists. It is notable that the IPPC's 1995 *Summary for Policymakers* says the following:

1. Greenhouse gas concentrations have continued to increase;
2. anthropogenic aerosols tend to produce negative radiative forcings (cooling);
3. climate has changed over the last century;
4. the balance of evidence suggests a discernible human influence on global climate;
5. climate is expected to continue to change in the future; and
6. there are still many uncertainties to be investigated.

A Closer Look at the 1995 Intergovernmental Panel on Climate Change

The 1995 IPCC *Second Assessment* clearly says that considerable progress has been made in the field of climate change since the 1990 assessment. This is in no small part the consequence of the concerted attention of the scientific community to the problem over that period. Climate change had come to the forefront of scientific investigation in the 1990s, and by 1995 the IPCC was able to state clearly its position in six important statements to policymakers. Their detailed findings included:

1. *Greenhouse gas concentrations have continued to increase.* Atmospheric concentrations of carbon dioxide, methane, and nitrous oxide

have risen since pre-industrial times by 30 percent, 145 percent, and 15 percent, respectively. If allowed to continue at current rates, carbon dioxide will be double (500 ppmv) pre-industrial levels by the end of the twenty-first century. Many greenhouse gases, such as carbon dioxide and nitrous oxide, remain in the atmosphere for decades or even centuries, and they do have a positive effect on radiative forcing of climate (leading to a warming of the Earth's surface).

Other greenhouse gases, such as chlorofluorcarbons (though not hydrochlorofluorocarbons) have remained constant in concentration because of the impact of the Montreal Protocol. It was signed only some fourteen years ago (in 1987), and is still having an impact. The effect of these substances on stratospheric ozone depletion will decrease substantially by 2050. Tropospheric ozone levels have increased in the Northern Hemisphere since pre-industrial times, potentially leading to more positive radiative forcing, although the trend over the past decade has slowed significantly.

2. *Anthropogenic aerosols, such as sulfates from fossil fuel burning, tend to produce cooling.* Aerosols resulting from human activities, such as combustion of fossil fuels and forest fire burning, do not reside long in the atmosphere and therefore tend to have a more localized effect than greenhouse gases. However, depending on their concentration, duration of source, and region of emission, the negative radiative forcing resulting from aerosols can sometimes be substantial enough to counteract positive radiative forcing by greenhouse gases.

3. *Climate has changed over the last century.* The IPCC report acknowledges that year-to-year variations in weather can be significant. However, examination of instrumental records has revealed some important changes that are believed to be systemic, and therefore not prone to misinterpretation because of spatial or temporal variability. Global mean surface air temperature has increased between .54 and 1.08 degrees Fahrenheit (.3 and .6 degrees Celsius) since the late nineteenth century and recent years have been among the warmest since A.D. 1860. (The change has been reflected more in nighttime than daytime temperatures.)

The change varies with region, being greatest over mid-latitude continents in winter and spring, while some areas, such as the North Atlantic Ocean, have experienced cooling. Winter precipitation has also increased in high latitudes of the Northern Hemisphere. Over the past 100 years, global sea level has risen between 4 and 10 inches (10 and 25 centimeters).

Finally, the warm phase of the El Niño Southern Oscillation, which leads to extremes in precipitation and temperature on a global scale, has been unusually persistent for the period 1990 to 1995 relative to the last 120 years (and through most of the remainder of the 1990s, although this was not known at the time of the 1995 report).

4. *The balance of evidence suggests a discernible human influence on global climate.* The IPCC document describes how variable natural climate can be, as a consequence of volcanic activity and solar variability, making detection of human influences on climate far more complicated. However, based on available data, twentieth century global mean temperature is at least as warm as any century since A.D. 1400 (the onset of the Little Ice Age). The observed record and climate models suggest it is unlikely that the current warming trend is purely natural. However, because the human influence on climate is still emerging as a signal from the noise of the natural trend, there are uncertainties in key factors such as magnitude and pattern of change produced by the interaction of natural and human-forced climate change.

5. *Climate is expected to continue to change in the future.* The best estimate from models of climate change that incorporate a mid-range greenhouse gas emission assumption is that global mean surface air temperature will rise about 3.6 degrees Fahrenheit (2 degrees Celsius) between 1990 and 2100. This is a third lower than the best estimate from 1990, because it assumes lower greenhouse gas emissions, the countering influence of sulfate aerosols, and more realistic views of how carbon dioxide interacts within the Earth's carbon cycle. Utilizing the lowest IPCC emission scenario, the increase in temperature is closer to 1.8 degrees Fahrenheit (1 degree Celsius). Regional temperature change could vary greatly from the global estimate. General warming is expected to result in an increase in the frequency of extremely hot days and a decrease in extremely cold days. Estimates of sea level rise because of thermal expansion of the ocean, resulting from increased global temperature, range from inches to 20 to 35 inches (50 to 90 centimeters) depending on the emission scenario that is chosen.

All models suggest that surface warming will be greater over the continents than the ocean in winter, maximum surface warming will occur in high northern latitudes in winter, minimal surface warming will occur over the Arctic in winter, and precipitation will increase in high latitudes during winter as the overall exchange in Earth's water balance becomes more intense (leading to more extremes in floods and droughts in some regions).

Most model simulations suggest a reduction in the strength of the conveyor (see figure 3.13) that transports heat from North America to Europe over the North Atlantic and a decrease in the daily range of temperature. Furthermore, most models reveal regional differences as a consequence of the effects of anthropogenic source aerosols.

6. *There are still many uncertainties to be considered.* It is not yet clear what future emissions will be or how the combined biological-geological-chemical system will respond to changes. Simulations in models of the effects of clouds, oceans, sea ice, and vegetation require further study. Longer instrumental and proxy records that represent the response and controls on the climate system must be evaluated to provide sufficient perspective from which to model and assess climate change. Surprises in the climate system, such as changes in thermohaline circulation in the North Atlantic, or ecosystem influences, are complicated and difficult to predict (IPCC, 1995).

In the United States, the debate over climate change often polarizes around the issue of whether global, national, and regional warming is a natural phenomenon or anthropogenic in origin. At first glance, this would seem to be a purely scientific debate, which is how we have treated it in this book so far. To settle the debate, science would look at the most robust and extensive records available and ask straightforward questions, such as:

1. Do we have evidence of warming or cooling trends in the past, before human civilization might have been a factor?
2. Are these warming or cooling trends more or less equivalent to what we observe today?
3. If the trends are equivalent, then a case can be made that natural factors are at work. If there is a radical difference today from anything seen before, a case can be made that the causes are anthropogenic.

Having gone through a process much like the one described above, and using the very complete records from Greenland, Antarctica, and elsewhere as our baseline, we have concluded in this book that the trends being seen today are indeed unique, and are most likely caused by human factors. This is not to say that we have not seen warming and cooling shifts of dramatic proportions, as in the case of the Medieval Warm Period and Little Ice Age transition, but the record on greenhouse gas emissions in the past century is undeniable, and clearly anthropogenic in origin.

We have also noted that, absent some unique factor, we should clearly still be in the Little Ice Age, rather than experiencing temperature increases. We conclude that the "unique factor" at work is humanity.

However, contrary to popular belief, science is not able to provide absolute, final answers to policy-related questions, at least not in most instances. Because science depends on verifiable data to draw conclusions and make suppositions, one can always argue that more data is needed—and it always is. The problem is that gathering data requires time and resources, both of which may be in short supply. As we have seen, retrieving the Ice Chronicles from Greenland took years and cost the federal government $25 million—and GISP2 is only one of the many climate change research projects that have been undertaken at public expense in recent years. Fortunately, subsequent ice coring efforts have cost significantly less, because we were able to develop methodologies through GISP2 that increase scientific and logistic efficiency.

It logically follows that the cost of obtaining more information to inform the climate change debate must be taken into account in deciding whether to wait for that information before taking action (not to mention the potential costs in human and economic terms if preventive measures are delayed that might improve the situation).

In the case of global warming, those who feel comfortable having government take action to protect the public good have a tendency to accept the evidence that climate change is being caused primarily by human beings and it is therefore incumbent on us to do something about it. These advocates may resist waiting for more science to be done, arguing that time is short, and we must take action *now*. In the case of the most recent report issued by the United States government on regional climate change, those who hold this position are inclined to focus on the document's descriptions of dire consequences. It is important to act on these results, but it is also essential to realize that further research will both clarify existing questions and also open up new avenues of investigation. It is the nature and the value of science to continue to explore.

Those in favor of limited government, and those whose self-interest will be affected by government action to curb global warming, are far more likely to argue that "we don't know enough yet," and that "these changes may well be caused by natural forces over which we have no control." These advocates are inclined to call for additional scientific research before making new policies. Their concern is that government regulation will be imposed to solve a problem that doesn't exist. In the case of the report on regional climate, they are naturally going to suggest that the report is only an approximation, and that it ought to consider "winners" as well as "losers" if climate change does take place.

This debate is reminiscent of the struggle over U.S. space policy in the 1950s. After the Soviet Union launched Sputnik, President Eisenhower came under pressure to create a national space agency and national space effort to catch up with the Soviets. He resisted such a step, because he feared that competing with the Soviet Union in that way would "Sovietize" the United States by giving the federal government enormous new and unprecedented powers. "Ike" believed that we might win the "race in space," only to lose it on Earth by becoming more like our opponents. He downplayed the importance of Sputnik, and argued for staying on the course that had already been laid out, which focused on minimal government intrusion onto the space frontier.

Opponents, many of whom already favored far greater government intervention in national life, especially the economy, seized on the space race to advocate massive government spending and a national space agency (which eventually became NASA). Led by Speaker of the House Lyndon B. Johnson, this group ultimately prevailed (McDougall, 1997). Presidential candidate John F. Kennedy used the so-called "missile gap" between the United States and Soviet Union as a campaign issue, and committed the nation to the moon landing not long after his inauguration. President Kennedy's interest in the moon race was driven perhaps less by an interest in space exploration, and more by a perceived need for a project in space that would overwhelm the Soviets on Earth, where an intense Cold War struggle was under way.

With hindsight, it is difficult to say which approach was better. Today, the United States and Russia are cooperating on the space frontier, and there has also been a powerful swing away from government domination of space enterprise, with private interests coming to the fore. However, regardless of which policy would have been better, it is instructive to see the extent to which politics infused the policy debate of that time.

The belief that human behavior is at the root of climate change appears to be held more widely abroad than in the United States, and the fight is increasingly over who will bear the burden of doing something about it. The less-developed countries argue that they had little to do with creating the problem, because the industrializing West put the pollutants in the air that are exacerbating the greenhouse effect. These countries want the benefits of economic growth that have been enjoyed for some time now in the more-developed world, and if that is going to create environmental problems, they want the West to find a solution—by cutting back on their own greenhouse gas emissions, for example.

Still, plans for modernization in less-industrialized countries may require revision, or at least reconsideration, regardless of what the more-advanced nations agree to do. If mega-nations such as the People's Republic of China were to follow the same path that brought Europe and North America into the industrial age, the impact on the global environment and climate could be unprecedented (McElroy, 1997). At the same time, when seen from the perspective of the Chinese leaders and people, it is unfair to ask them to remain in poverty when others are enjoying a more pleasant lifestyle.

As one reporter noted in the run-up to the Kyoto climate change conference in December 1997:

> Probably the most contentious question to be faced in Kyoto is whether all countries—rich, industrialized nations such as the United States as well as poorer but fast-growing countries in Asia and South America—will be required to limit their use of fuels to meet emission reduction goals. (Howe, *Boston Globe*, 1997)

On December 10, 1997, representatives from more than one hundred countries who had gathered in Kyoto, Japan, agreed to begin controlling emissions as a move toward reducing global warming. However, the treaty was not ratified by the U.S. Senate, because of concerns about the economic impact, and questions of whether the Kyoto treaty distributed the burdens of compliance equally (McGrory, *Boston Globe*, 1997).

It is also a sign of the times that disagreements over environmental policy began to emerge in the 1990s at meetings of world leaders that had previously focused on economic and political issues. For example, in June 1997, the leaders of the seven most industrialized nations and Russia met in Denver. Most of the attendees wanted the United States to agree to a 15 percent reduction in greenhouse emissions by 2010. President Clinton had previously resisted specific targets, which drew criticism from some of those attending and observing the meeting (Warrick, *Boston Globe*, 1997).

As at this "G-8 summit," the pressure for action falls increasingly on the United States, with Europe often going along with the less-developed countries. For example, at the United Nations Earth Summit, also held in 1997:

A rift over air pollution clouded the start yesterday of the second United Nations Earth Summit, as European governments criticized the United States for failing to commit to a timetable for reducing the emission of gases believed responsible for global warming . . . Vice President Al Gore, welcoming delegates to the summit, said the United States would pursue a treaty that would give governments flexibility in deciding how to achieve reduction in pollution. But he warned that failing to reduce the so-called greenhouse emissions could result in global environmental disaster. "Unless we change course, during the lives of our grandchildren, concentrations of gases will reach levels not seen on this planet for more than 50 million years," Gore said. (Warrick, *Boston Globe,* 1997)

In Gore's speech, we see an interesting phenomenon, which is that of a political figure acknowledging the potential danger of the situation, while searching for a solution that will meet the needs of those who oppose massive government action.

As we also noted earlier, while the climate change challenge divides nations, it also provides a powerful impetus to unite in the face of a common peril. In his first speech before the United Nations as newly elected prime minister of Great Britain, Tony Blair spoke to the same Earth Summit conference and said, "We are all in this together . . . No country can opt out of global warming or fence in its own private climate" (Warrick, *Boston Globe,* 1997).

Blair is certainly right. While there may well be regional climates and regional climate change, all of these are part of a single Earth system. This means that while policies may vary by country, they will not be very effective if they are not coordinated by some global consensus.

At the same time, the marriage of science, policy, and politics that we discussed in the previous chapter is often an uneasy union. In a newspaper article published just before the Kyoto conference, scientists bemoaned the dangers that were being posed to their ethics by the intense debate over global warming. Kevin Trenberth, a climate researcher at the National Center for Atmospheric Research in Colorado, said, "If you try to be responsible, then nobody listens . . . The rhetoric is very strong. The environmentalists tend to exaggerate on one side, and the skeptics exaggerate on the other side" (Chandler, *Boston Globe,* 1997a).

In a similar vein, Patrick Michaels, a climatologist at the University of Virginia who has questioned higher estimates of warming that may be experienced in the near future, told a story of being on a television program with a representative of the Environmental Defense Fund. When the taping was over, the producer reportedly complained that the participants weren't "disagreeing enough" (Chandler, *Boston Globe,* 1997a).

The media focuses on reporting controversy for the pragmatic reason that disagreement draws an audience, but also because that is an appropriate function the media alone can serve in helping citizens sort out issues. The press is drawn to climate change precisely because it affects everyone, and because there is real disagreement about its causes and consequences.

Confronting the Choices

While the debate over causal factors and responsibilities for change can be complicated, it doesn't take a climate scientist or an expert in public policy to anticipate the options for the world. It simply requires a bit of logical thinking. Looking ahead to the future, we are faced with a limited range of choices. These are:

1. Make no changes in the human social and economic systems, and try to use technology to redirect the climate system, thereby correcting the effects of our own behavior;

2. Make no changes in human social and economic systems, except to adapt to shifts in the climate system caused by phenomena such as global warming;

3. Modify the behavior of human systems to reduce our impact on the natural system, allowing it to revert to something more like its normal state, thereby to become more predictable;

4. Choose the best combination of the above options that can be agreed upon by the global community, and the approach that will have the greatest positive impact.

Let's now consider each of these options in turn in light of the science that has emerged in the past decade through projects such as GISP2.

Modifying the climate system. Having used science and technology to influence climate in a direction that we do not like, we might now look to the same tools to control climate in ways more compatible with our wishes.

The dream of controlling, rather than adapting to climate and weather, is an old one, and it dies hard. For example, cloud seeding to create rain is an on-again, off-again approach to climate control that has never been fully successful, but has never quite gone out of fashion, either. To take just one example: In a 1983 *Science Digest* article, optimistic analyses of future abilities to *predict* the weather led to more problematic *control* scenarios, with statements such as, "Once they understand the weather, meteorologists hope that they may someday be able to dictate its course—or at least modify some of its more destructive aspects" (Mann, *Science Digest,* 1983).

This article was written some two decades ago, and we still are not very close to humanity's dream of controlling the weather, not to mention climate. Nor does this kind of thinking address the question of how changing weather patterns on one part of the planet for the better might affect weather patterns elsewhere, possibly for the worse.

In the words of H. H. Lamb, "father of modern climatology":

> The more grandiose schemes for "altering the face of nature"—plans such as the diversion of the Gulf Stream or the Siberian rivers or abolition of the Arctic ice—should be approached not only with caution but with skepticism. As long as our capacity for forecasting the weather is limited and sometimes marred by gross errors affecting large areas, our ability to foresee the consequences of any deliberate manipulation of the climate system that might be attempted are subject to the same danger. (Lamb, 1982)

We can now add to Lamb's comments the realization that climate is controlled by many factors. These factors need not all operate in the same way, at the same time, and they need not always produce a nice linear one-to-one response. Acting together, the smallest additional increase in a climate control or the addition of a new climate control, even small, could be the proverbial "straw that broke the camel's back." In science, we think of the jump from one style of response, such as linear, to another, such as non-linear, as a "threshold effect." We ask, then, when has the system been pushed too far?

Lamb's comments make sense, but that doesn't mean that everyone will listen, nor that efforts to manipulate the weather and climate will cease.

Still, the idea that it will be acceptable to allow the climate situation to continue as it is, while working to find a technological "fix" for situations such as global warming, belies the undeniable complexity of the climate system. It also ignores the time that will be required to implement the imagined solution.

In the case of greenhouse gases, most of the policy options that have been seriously considered so far involve reducing future emissions, and we do not yet have a technology that will safely remove existing gases from the atmosphere. In the time it would take to develop that capability, greenhouse gas levels will continue to rise, with nothing being done about what is already present in the atmosphere.

At the same time, as a newspaper article on global warming points out: Even under the most draconian proposals for limiting greenhouse gas emissions, the buildup in the atmosphere is expected to continue until it is at least double the pre-industrial level—and only then will level off (Chandler, *Boston Globe*, 1997b).

Given this harsh reality, the argument for natural or "low-tech" ways of reducing the impact of greenhouse gas buildup in the atmosphere gains credence. For example, planting more trees is one idea with some merit, since plants do absorb carbon dioxide, and tree planting would have no negative side effects that anyone can imagine.

Other ideas that follow the same pattern would include the following:

1. *Changing the fuel-to-air mix on jet aircraft.* This simple step would increase the amount of microscopic particles in the exhaust, putting more water vapor into the atmosphere, and in effect, creating more cloudiness, which has a cooling effect. But there is a trade-off in the form of increased pollutant levels.

2. *Helping more plankton to grow.* The oceans are a sink for CO_2, because plankton absorb it. Scientists have already demonstrated that lack of iron is the primary restraint on plankton growth, and that putting iron dust into the ocean will cause them to grow rapidly, thereby absorbing more CO_2. Further, as phytoplankton increase in number, they give off sulfur gases that shield incoming radiation and cool the lower atmosphere.

3. *Increasing reflectivity.* It is already known that any increase in natural albedo, or reflectivity of the Earth's surface, can lead to cooling. Slight changes in construction methods might have a similar effect. For example, we could add sand or ground glass to asphalt to increase its reflectivity. Given the amount of paved roads worldwide, this could have significant impact. But this strategy also has potential effects on local vegetation and the melting capacity of snow- and ice-covered roads.

Given the potential seriousness of global warming, it would be difficult to criticize anyone with a good idea for mitigating its effects. However, many, with some justification, may fear that speculation of this kind will distract us from making good policy and taking other steps that are necessary in learning to live with climate change. Considering these and other alternatives makes good sense, but it would be better not to have to "fix" things at all. We can only avoid this necessity by attacking the root of the problem.

Adapting to changes in the climate system. Adaptation to changes in the climate system is another option, and we are already seeing a preview of that approach. For example, there is growing opposition to government flood insurance programs that encourage people to build homes too close to the ocean, thereby running the risk of being inundated by rising seas.

If we were to make no changes in human behavior and are not successful in making major breakthroughs in modifying weather and climate, air and water quality will continue to deteriorate and climate will become more of a wild card. The relatively stable climate on which civilization's evolution has depended will become increasingly volatile, and we will experience more of what scientists call "surprises." The only alternative will be to create a new kind of civilization that does not assume stable climate to be a foundation of its success.

What kind of civilization might this be? This is difficult to imagine, since examples of previously successful adaptation (the Inuits come to mind) seemed to depend on very simple social structures, using very little technology for survival. Today's global civilization is built on high technology and a complex interdependence.

Could such a culture respond to a major worldwide calamity, such as a rapid climate change event (RCCE) that is made even more powerful by human factors? Among the most intriguing possibilities is that human activity may reinforce an existing RCCE, or even trigger a new RCCE. But how might a new RCCE of this kind be catalyzed? As one example, if the current global warming scenarios are correct, and temperatures continue to increase in some regions, the high latitudes will experience melting of permafrost and glaciers that might flow into the North Atlantic. That in turn would increase the ocean's salinity and decrease the heat along the Gulf Stream to Europe. Ironically, that continent would become colder, while other regions would warm up.

In fact, human beings have the ability to influence nearly every variable in the climate change equation except solar output and volcanoes, and decisions made by humans can move us in the direction of predictability or uncertainty. At the moment, however, the system may be drifting into an unstable state that has never been seen in the past.

By virtue of adding new climate forcing agents we have made the situation more complex, so it is not surprising that the climate is more unstable. As we have noted elsewhere, it is ironic that just at the moment when humanity has created the ability to understand climate based on the Ice Chronicles and other records, civilization is also creating uncertainties that may cancel out the gains made in the predictive realm. These gains in prediction, in turn, would make the opportunities for adaptation more feasible.

This is not to say that it is impossible to model the increases in CO_2 or changes in albedo that current practices are likely to produce. The problem is that we simply have no way of knowing their impact with any degree of confidence.

As we have already noted, some argue that we should just take a laissez-faire approach to climate change. After all, there will be winners and losers if global warming continues. Presumably, there would also have been winners and losers if the cooling trend of 1940 to 1970 had gone on for a longer period. Some even believe that a warmer world will, on balance, be a better world. For example, Thomas Gale Moore, a senior fellow at Stanford's Hoover Institution, is quoted in *Global Change* magazine as saying that "warmer is better," and that "a rise in worldwide temperatures will go virtually unnoticed by inhabitants of the industrial countries" (*Global Change*, 1996).

Moore's contentions beg the question of whether people in *developing countries* will "notice" a rise in temperatures, or whether island nations will "notice" a rise in sea levels, but it also overlooks several other concerns. This argument that global warming is, on balance, a good thing, fails to address four key issues: First, current trends could very well be creating increased instability in climate. Second, current trends are having negative effects on the environment that are independent of climate: Air and water quality are deteriorating world-wide, and people are dying as a result of that decline. Third, a warmer global climate will likely result in much greater levels of disease in more parts of the world: Tropical climates generate more diseases than polar climates simply because disease-causing organisms are able to survive and spread more easily. It is difficult to see who the "winners" will be if diseases increase on a global scale. Fourth, climate change is not just a question of warmer or colder. We know from the Ice Chronicles that dramatic changes in wind speed, wind direction, and precipitation also characterize transformations in climate.

As just one example of this line of thinking, Rita R. Colwell, director of the National Science Foundation, writing in the December, 1996 issue of *Science,* notes that infectious diseases are already "resurging." She says that malaria, which currently kills about two million people globally, could take an additional million lives if global temperatures rise, thereby allowing the mosquitoes carrying malaria to expand their range. Colwell also reports on studies of cholera pandemics, demonstrating that cholera microbes hitch rides across the ocean on phytoplankton, and are more successful in moving from one region to another when sea surface temperatures rise (Colwell, 1996).

Modifying human systems. To date, this option has triggered the most political controversy and the greatest media attention, but the debate has often generated more "heat than light." A closer examination of various treaties is in order to show why some work well, and others work hardly at all. Such an analysis shows that changing human social and economic systems to ameliorate an environmental problem can be easier at some times than others. For example, the nations of the world met in Montreal and successfully created a treaty on phasing out substances that were contributing to the ozone hole.

The Montreal Protocol has worked for many reasons, not least of which was that reducing the production of chloroflurocarbons clearly was not going to bring the global economy to its knees. As well, few divisive issues of equity (in terms of which nations would bear the burdens) had to be resolved in negotiating the treaty.

Montreal represents a case in which scientists, politicians, and public policy experts came together to produce a relatively painless result, and one that has already produced real benefits. Greenhouse gas emissions are quite another story. As we pointed out earlier, today's global economy is still largely driven by fossil fuel consumption. Huge industries stand to gain or lose enormous amounts of money depending on how treaties are drawn up around the greenhouse gas emissions issue.

Policymakers have real concerns about the economic impact of worldwide regulations designed to limit emissions. There is also a thorny ethical debate over who should bear the brunt of the challenge—the developed nations, who have created much of the problem during the past few hundred years, or the developing nations, who will surely exacerbate the warming trends if they follow the same path of industrialization already taken by others.

The Kyoto conference on climate change, called to confront these matters, produced an ambiguous resolution of these difficult issues. On the one hand, 160 nations did agree to endorse a plan that included provisions for the United States to cut emissions 7 percent below 1990 levels, with the European Union and Japan agreeing to cut 8 and 6 percent respectively (Lakshmanan, *Boston Globe,* 1997).

However, the agreement did not bind the developing nations to any targets for reduced emissions, which led many U.S. senators to announce their immediate opposition to the treaty, using terms such as "fundamentally flawed," and "dead on arrival" (McGrory, *Boston Globe,* 1997).

With the details being left to yet another meeting, this much seemed clear in the fall of 1997: Some modification of global human systems, post-Kyoto, seems inevitable. The issue will be how the world community comes together to create an initial structure within which every nation can live comfortably and equitably.

✳ THE FUNCTION OF FEAR

People often ask me, "How can you do the things you do? Isn't your work really dangerous? Aren't you sometimes afraid? Doesn't your wife worry?" The answers are clear enough in my own mind, but I'm not sure they make sense to the people who ask me those kinds of questions. I tell them that, yes, my expeditions are indeed potentially dangerous, but we make every effort to prepare for every eventuality, and we reduce the danger factor as much as we can; we're scientists, not daredevils. That's why I haven't climbed to the summit of Everest, and don't plan to.

I also tell them that, yes, I'm often terrified. Ignoring warning signs of danger in the name of bravery could be the most dangerous behavior of all. Fear, if it doesn't paralyze you, has a positive function. It is your body and your mind saying to you, "Be careful, be very careful. You're doing something dangerous right now."

In the case of global climate change, the warning signs of danger are not as easy to detect as the immediate dangers that we face on an expedition. Climate change takes years, and few, if any, of us can remember from year to year the subtle alterations in weather that in sum can signal significant change. I tell people that, ultimately, the danger of our expeditions cannot be seen outside the context of the payoff. The potential for understanding climate change, which could have a dramatic impact on the way we live, cannot always be viewed over short time periods.

When people talk about "gathering more data" to make better policy, what they really mean is that someone has to go out (sometimes to inhospitable regions) to obtain that information. My colleagues and I are engaged in a process of discovery that we hope will have great benefits to all of humanity. If we can contribute to the problem of predicting climate more effectively, we may be able to contribute dramatically to the quality of life and the course of civilization. Current concerns about global warming and the need for information on which to base policy add real urgency to our work.

A hybrid approach. It's easy to polarize the options for responding to climate change, and it certainly makes for a better story when conflict rages between or among "sides" in the debate. This approach leads to the idea that the world should *either* adopt draconian measures to reduce greenhouse gas emissions *or* avoid all regulation, let nature take its course, and accept that there will be "winners" and "losers" as the climate changes.

In fact, precisely because so many interests need to be satisfied in developing a global climate policy, a combination of available policies is already emerging, and that may not only be the most *practical,* but also the most *desirable,* approach.

Certainly, we can take modest steps to modify the climate, such as planting more trees, which provide other benefits beyond their impact on the climate. I can see no obvious reasons not to pursue those projects. Similarly, being more adaptable to climate change is a positive, survival-oriented trait for societies, because our actions in the recent past are likely to affect the climate for some time in the future. For example, in commenting on the Kyoto agreement to reduce emissions by 2012, Michael Oppenheimer of the Environmental Defense Fund argued that even 7 percent cuts would only delay the doubling of carbon dioxide in the atmosphere by a couple of years (Chandler, *Boston Globe,* 1997c).

And as we have come to understand from our study of the Ice Chronicles, natural climate alone has the ability to transform our experience of the weather within a very short period of time. Therefore, we must abandon the long-held assumption that climate is going to remain a stable constant in our lives.

We would also be wise to modify how our human social and economic systems work, though we may find better and worse ways of doing so. Discharging pollutants into the atmosphere at the rate of the past century is likely to contribute not only to global warming, but also to the destruction of forests through acid rain, the death of large bodies of water, and an increase in the incidence of human disease. Policies that modify these important systems do not necessarily require elaborate global bureaucracies that stifle individual or corporate enterprise. On the contrary, we can probably discover approaches that provide free market incentives to support environmentally sound behavior, and that would, in any event, work better than regulation (Hahn and Stavins, 1999).

Summary

Science is an organized enterprise that attempts to reveal and explain the behavior of natural and humanly provoked phenomena. It is based on the scientific method, which involves the gathering of data, the testing of hypotheses against experience, and a commitment to finding the objective truth that underlies the behavior of these natural systems.

Politics is not a science, and it has often been called "the art of the possible." It is an effort to find those compromises and arrangements that will satisfy competing needs of different power groupings within societies and allow those societies to continue. While conflict remains a part of any political system, when the system works well, conflict is relegated to the domain of verbal disagreements, rather than escalating into violence.

Scientists can propose policies, as can academics and staff members of domestic or global administrative systems. However, policies that are adopted by the political system are often the product of the interplay between science and politics—a system that seeks ideal answers versus a system that devises realistic solutions.

As important as it may be, climate change policy is no different in this regard. The Ice Chronicles tell us a great deal about how climate has behaved in the past, and how it is likely to behave in the future. Scientists who interpret these natural records may well see an ideal path toward ameliorating global warming or reducing the impact of climate change. However, the policies that are put into place will always include a heavy dose of politics, just as they did in the eighteenth-century struggle between the ancestors of today's political parties.

One of the main purposes of this book is to ensure that such an important issue is addressed with the maximum amount of solid scientific information at hand. The only thing worse than ignoring good science is embracing bad science!

We live in a world in which concentrations of greenhouse gases, acids, and toxic metals have risen far above anything experienced since the beginning of human civilization and well before. Pristine views of the landscape are gone in all but the most remote parts of the planet, maintaining decent air and water quality requires constant vigilance, and the inevitability of climate change is becoming more obvious daily. Change will continue and, as the ancient Greeks advised us, "everything in moderation" may be the best compromise in our complicated world of politics, technology, and economics.

8

Learning to Live in a Changing World

For civilization as a whole, the faith that is so essential to restore the balance now missing in our relationship to the earth is the faith that we do have a future. We can believe in that future and work to achieve it and preserve it, or we can whirl blindly on, behaving as if one day there will be no children to inherit our legacy. The choice is ours; the earth is in the balance.   —Al Gore, *Earth in the Balance*, 1992

Human evolution on this planet is going to become increasingly connected with good environmental policymaking, and it is now clear that this process will be global in nature, an undertaking for the entire human family. Globalization has already claimed the allegiance of many human institutions, such as finance, yet policy is still made by sovereign states, each with its own agenda and understanding of national self-interest. Still, recent initiatives offer solid evidence that society can change its ways when enough scientific evidence of real danger is available.

The Montreal Protocol on substances that deplete the ozone layer was signed originally in 1987 and amended in 1990 and 1992. It is a good example of a global response to a global problem. All of the signatory nations made a firm resolution to significantly abate the cause of stratospheric ozone depletion, and it worked. On the other hand, while almost no one expressed total satisfaction with the treaty on greenhouse gases signed in Kyoto, the existence of any kind of treaty is a step in the right direction. In the United States, the Clean Air and Clean Water Acts passed in the early 1970s have had a measurable impact, their effects manifesting in ice core records and through monitoring experiments.

These examples demonstrate that national and international action can indeed counter the impact of the human social system on the natural climate system, and that the results will sometimes show up within a short time period.

Moreover, the disagreement over the temperature effects potentially produced by global warming is not as significant as some would have us believe. The so-called "cynics" who tend to downplay the impact of global warming expect a 2.7 degree Fahrenheit (1.5 degree Celsius) increase in temperatures worldwide, while the 1995 U.N. International Panel on Climate Change report suggested a range of 2.7 to 8.1 degrees Fahrenheit (1.5 to 4.5 degrees Celsius). At the top of one range and the bottom of the other, the two camps are actually in agreement.

As we enter a new millennium, then, we must confront a great responsibility to future generations to make the right choices. We must ask ourselves, "What expectations about policy and the environment are realistic; what do the Ice Chronicles tell us about the effects of policy on quality of life and climatic conditions? Even if we have no final answers, are there some basic principles to guide us?"

The answer is, "Yes, there are." We now know that human civilization has probably pushed the natural system beyond the point where we can find any easy solutions, but real improvements can still be realized. The best policy will be a combination of the options detailed in the previous chapter—that is, using creative means to reduce greenhouse gases, adapting to climate changes that cannot be ameliorated, and shifting the human economic and social systems' priorities toward climate-stabilizing technologies to drive the new economy.

We must also avoid the mistake of identifying climate change and environmental quality with one another. Protecting the environment is a more comprehensive issue than climate change, just as climate change is a larger concern than global warming. Focusing only on warming and cooling distorts the central insight of the Ice Chronicles, which is that climate change—including warming and cooling—is natural, but accelerated greenhouse gas emissions are destabilizing the climate beyond the ranges of natural climatic variations.

Therefore, our global policies really ought to focus on actions that lead to greater climate stability and predictability while also enhancing environmental health and well-being, not simply on affecting global temperatures. Policies must also be considered in terms of the difference between long-term and short-term impact and effects on the climate and on human social institutions.

✳ THE ENDURANCE OF SIR ERNEST SHACKLETON

We owe a lot to the explorers who went before us. They probably couldn't have imagined that their adventures would become so important to our understanding of ecology and of the delicate balance we must maintain to live successfully on the Earth. But their contributions have been enormous.

When I think of role models among the great explorers, Sir Ernest Shackleton always comes to mind. One of his finest qualities was an ability to set reasonable goals. He saw his work as contributing to the twin causes of exploration and science, and within that context, he focused on the most important priorities in every expedition that he undertook.

Even with his clarity of intent, however, Shackleton often found himself in the difficult situation of deciding between going forward to explore the unknown and making sure his party survived. The open boat journey from Elephant Island to South Georgia made by Shackleton and his colleagues in order to gather a relief ship to save

FIGURE 8.1. Examination of coastal mountain glaciers like these on South Georgia's Bay of Isles provide a sensitive monitor of change in sea surface and air temperature. *Photo by Paul Andrew Mayewski (1976).*

his ill-fated expedition is certainly one of the great epics of the polar expedition era. In every instance, he chose to save his people, on the assumption that it was better to come back again than to lose the life of even one person.

In Shackleton's time, those kinds of decisions made a big difference, because it took so much longer to get to Antarctica than it does today. In his entire career, he only went to Antarctica three or four times, whereas I've been fortunate enough to go there fifteen times, and I will probably have many more opportunities in the coming years.

I visited South Georgia in 1976, where I came upon Shackleton's grave. He was buried on the island after he died making yet another attempt to penetrate the Antarctic. South Georgia is a spectacular setting, with grasses and abundant wildlife near the shore and ice-covered mountains. Formerly a whaling station, it is now the home of a British scientific base. As I walked along the shores and up into the mountains, I could not help but think what a beautiful and peaceful

FIGURE 8.2. Tents and clothing styles have not necessarily changed since the turn of the nineteenth century, when exploration first started in Antarctica, despite the presence of many modern innovations like heavy lift ski-equipped C-130 aircraft and satellite communications. Double-walled Scott tents and loose windproof over-clothing are still used. *Photo by Paul Andrew Mayewski (1989), 40 kilometers from the South Pole.*

place it was. For those who like remote, sub-Antarctic regions it would be a perfect home! I thought it was also perfect because it was (I believed) protected from all of the politics and violence of the rest of the world.

A few years later, of course, the Falklands War broke out between Argentina and Britain, and the island was attacked by the Argentines. Even the remotest parts of the planet are vulnerable.

Immediate versus Long-term Effects

Changes in human social behavior can have immediate impact on some aspects of environmental deterioration. For example, acid rain is a short-term cause-and-effect environmental process. Acid precipitation (the more accurate term for "acid rain") would stop tomorrow if those activities causing it were to cease.

Toxic trace metals would also disappear quickly from the environment once we stopped emitting these substances. However, even after emissions of these substances have ceased, they can still be remobilized into air and water. A classic example can be found close to my former home, in a large estuary called Great Bay, in New Hampshire. Decades of tannery and clothing mill activity led to extremely high levels of toxic metals in Great Bay. In the 1970s and 1980s, the mills disappeared and regulations controlling the release of waste water from homes and industry were enacted.

The water in Great Bay became cleaner and cleaner. Tidal exchange between the bay and the ocean led to further cleansing of the system and today Great Bay is a beautiful sanctuary for wildlife. However, buried within the sediments of Great Bay are the toxic metals that once nearly killed it. Dredging, changes in erosion by ocean or rivers, or increased turbulence of the water could all lead to the re-incorporation of these toxic metals into the bay. The problem lies buried like so many environmental hazards caused by human activity.

Even with successful environmental policies such as the Montreal Protocol, the fix is not always immediate. While we know how to halt the destruction of the ozone layer, and that important effort has already begun, restoring the layer will require 70 to 100 years, even if some equilibrium is reached in our own time. The reality of this time lag means that humans will have to be extremely careful about exposing themselves to the sun throughout the next century, and we are likely to see many new cases of skin cancer over that period. The sooner we start the sooner we reap the benefits.

With greenhouse gases, the best that we can accomplish right away is to arrest the problem, but we cannot be assured of seeing any major changes in the near term. When it comes to the buildup of these gases, the planet is already in an unprecedented situation. Changes in policy will only ensure that any further deterioration is not the result of new human activities.

Recalling the various options discussed earlier, it seems clear that humanity will need to make some adaptation to living in a world in which at least some regions are warmer than normal. Even if every automobile were banned from the roads today, adaptation would still be required because there are many other sources of pollution. We have other good reasons for making changes as soon as possible, however. We know from the records that air quality is much worse today than it was a few decades ago. The battle to preserve clean air depends on solid legislation and constant vigilance. Improving air quality by reducing emissions is a worthy goal, even if we cannot avoid the uncertainties of a much less stable climate in the short term.

In the future, the human responses to climate changes will vary, as they have in the past, and may well produce results that are not altogether negative. For example, some of the greatest of ancient civilizations, such as the Egyptians along the Nile, and the Indians along the Indus, emerged in response to "negative" climatic conditions, that is, severe drought in other parts of those countries. We may well find that the challenge of climate change will produce stronger human societies in the long term, but at what cost and to whom?

In extreme regions such as Greenland (the Norse example) or western Asia (the Mesopotamian example), small changes in temperature or precipitation, respectively, can produce dramatic changes in climate and in civilizations that depend for their survival on certain conditions. Outright collapse is not always the result, but radical readjustments have been quite common. Human beings are certainly going to do whatever they can to avoid collapse of today's fledgling global civilization, and on that score, there really are no "sides," as such. In fact, much of the debate that is now under way between those who emphasize protecting the environment and those who focus more on economic growth is about whether we have begun to prepare ourselves sufficiently for a radical readjustment, and indeed whether such an adjustment is really necessary.

In a New York *Times* article that presaged the Kyoto summit, William K. Stevens wrote:

> With President Clinton having accepted the dominant view among scientists that global warming is a serious matter, the increasingly urgent debate over what to do about it has largely shifted to the question of how restricting emissions of heat-trapping greenhouse gases may affect the nation's economy. And there, perplexity reigns. (Stevens, New York *Times*, 1997)

In his article, Stevens went on to describe the various measures under discussion, such as promoting alternative energy sources, imposing a "carbon tax," and allowing "emissions trading."

Many of the concerns expressed about over-regulation to achieve environmental protection may perhaps best be met with solutions such as emissions trading, which caps global emissions at a specific level, then allows companies (or countries) that have been successful at reducing greenhouse gases to sell their emissions permits to those that are having problems.

Assuming that there will be no quick fix, and that some adaptation may be required to cope with the future results of past actions, policies should not be too draconian. Instead, new ideas that support voluntary behavior moving in the right direction should be given a hearing.

Predicting climate change is difficult, but predicting the effect of policy decisions on human social systems is next to impossible. In the aftermath of the Kyoto conference, expectations of the impact of the agreement ranged from "zip" by Amory Lovins of the Rocky Mountain Institute, to the assertion by William O'Keefe of the industry-funded Global Climate Coalition that the treaty would "slash U.S. economic growth in half" (David L. Chandler, *Boston Globe*, 1997).

With all of these uncertainties, the global consensus that has developed on climate change is probably not going to allow the continuation of unfettered economic growth based on fossil fuel consumption. Moreover, while national sovereignty will most likely survive the climate change debate, nations may be forced to surrender more freedom of action than they would like, as the notion of seeking the common good over the national good takes hold.

☰ TWENTY-FIRST CENTURY SCIENCE

I'll never forget the evening in the fall of 1999 at the Museum of Science in Boston. The occasion was a festive send-off of the International Trans Antarctic Scientific Expedition (ITASE). The event featured expedition members walking through the museum and chatting with the crowd while wearing the kinds of clothing that we wear in Antarctica. Our tents and equipment were on display as well.

Led by the director of the Museum of Science, David Ellis, we raised a champagne toast at the end of the evening. I felt a thrill not unlike what I had experienced when I found out that we had the funding for GISP2. I have been on many Antarctic expeditions now, but I hadn't looked forward to leading any of them more than the ITASE, a multinational project that is undertaking a series of crossings of the Antarctic continent. We are doing this because, with all our explorations of Antarctica, we still know relatively little about it. That's partly because it is such a huge place, and it takes time to learn all there is to know.

FIGURE 8.3. Tucker SnoCats pulling equipment across the West Antarctic ice sheet during the first season of the United States' contribution to the International Trans Antarctic Scientific Expedition (ITASE). *Photo by Paul Andrew Mayewski (1999).*

For a long time, we didn't understand how environmentally sensitive Antarctica is, for example. We've found evidence of radioactive fallout and the first hints of toxic metals in Antarctica, far from their sources. We would never have imagined that the ozone hole would first be detected there, and I wonder what else will eventually be seen for the very first time on its frozen wastes.

ITASE will take ten years and will involve fifteen countries. We are traveling in groups of five to fifteen people at a pace of 1,000 kilometers per season, using snowmobiles plus large oversnow vehicles. We are collecting ice cores along the route, and taking four to ten weeks to complete our transits, depending on the severity of the storms we encounter. We will be drilling back in time 200-plus years per site to develop an even more complete record of climate change on the Antarctic continent.

The way I like to describe ITASE is that we are using nineteenth-century science with twentieth-century tools to solve twenty-first-century problems. It's a global enterprise to address global problems, in the spirit of IGY and other similar programs undertaken by the international community during the past forty years. I love doing the work and hope it will prove to be as useful as our previous research in Greenland and elsewhere.

But the obvious question arises: Why go to such a remote place to contribute to our understanding of environmental change?

Antarctica is a vast continent, close to one and one-half times the size of the United States, with an average ice thickness of close to 6,000 feet (1,829 meters). It is the world's greatest storehouse of ice cores. Our goal is to understand how climate and the chemistry of the atmosphere has changed over Antarctica from a time prior to the industrial revolution to the present. Although it was the first region to suffer the effects of an expanding ozone hole, Antarctica still has the most pristine atmosphere in the world.

It can, in effect, act like the proverbial canary in the mine, warning of problems we cannot yet see. When Antarctica begins to show marked increases in toxic metals, acids, and other pollutants, the rest of the atmosphere must surely be heavily saturated with these substances.

In addition to offering a storehouse of ice cores and having a pristine environment, Antarctica also has the following links to the rest of the world. It is:

- *The largest storehouse of freshwater on the planet.*
  Seventy percent or more of the world's freshwater is tied up as Antarctic ice. It is also the cleanest freshwater on Earth.
- *A major site for the production of the cold deep water that drives ocean circulation.*
  The only Northern Hemisphere production site is the North Atlantic.
- *A potential contribution to sea level.*
  Past records reveal that during the last interglacial episode (about 118,000 to 135,000 years ago) temperatures were 2 degrees Celsius warmer than today and sea level close to 15 feet (4.6 meters) higher. Portions of Antarctica are believed to have melted at this time, raising sea level.
- *A major player in Earth's albedo dynamics.*
  The vast white continent reflects nearly all incoming radiation. The white area almost doubles during the Antarctic winter, when sea ice surrounds the continent, changing the ocean from dark blue-green to white.
- *An important driving component for atmospheric circulation.*
  Cold winds stream off the high interior of Antarctica over smooth slopes, affecting climate as far north as the Himalayas.
- *Encircled by the world's most biologically productive oceans.*
  Cold, nutrient-rich waters surrounding Antarctica are the home to abundant marine life.

Principles for a Positive Future

The environmental community has focused much of its attention recently on "sustainability." This is a useful term because it encompasses several different ideas in a single word. Simply put, sustainability is concerned with how to contain the dual goals of economic growth and environmental protection within the social and political realities of the global community.

Those concerned with sustainability assume that economic development and environmental protection are both good in and of themselves, and that they need not be in conflict with one another. The challenge is to use human ingenuity to balance the two in constructing realistic policy alternatives.

We believe that climate change is a subset of the discussion about environmental protection, and that sustainability comes into play when we make policy about climate change. But can we identify basic principles on which to base a sustainable policy regarding climate change? We believe we can, and submit the following ten "principles for a positive future" as part of the world conversation on sustainability.

Unlike the Ten Commandments that were recorded in stone and have served as a moral and ethical code for millennia, these ten principles were written even earlier in the ice sheets and glaciers of Greenland, Antarctica, and the Himalayas. Their value, as compared with the commandments, remains to be proven by future practice.

Principle 1. Climate change will have both positive and negative consequences. We should certainly be concerned about climate change that creates instability, destruction, and death, but we should not overstate its potential negative impact on society. Human beings are flexible and resilient, and history shows that climate change can function as a spur to social evolution and positive changes in civilizations. In fact, it is possible that civilizations might not evolve, because they would not be subjected to significant challenges, if climate remained absolutely stable over very long periods of time. Further, without climate change some areas would not be habitable. Climate change may be important, in ways that we have not yet fully understood, to human social evolution. Social climatology is a field that needs much more attention and development.

Principle 2. We should not wait for perfect knowledge to take needed actions. No scientist would ever resist gathering new data in order to better understand the natural world—information gathering is in the very nature of the scientific enterprise. However, the research that has already been invested in climate change has paid off, and it should be used. Reading the Ice Chronicles tells us that unprecedented levels and rates of change in certain factors such as CO_2 concentrations are clearly correlated with human activities.

We will always need to know more, and we should avoid shortsighted decisions to stop monitoring projects, such as occurred after the 1957 to 1958 International Geophysical Year experiments when monitoring of ozone and CO_2 was almost terminated. However, we do know enough now to justify taking clear and determined actions.

Principle 3. We must set a long-range climate change policy and then be patient. While we have sufficient scientific data in hand to justify changes in public policy now, these steps should not be taken with the expectation that immediate results will follow. As just one example, even if all of the automobiles were taken off all the world's roads today (and no one is proposing that this should happen), we would have to wait for many years for the effects to be felt. More modest plans will clearly take even longer to produce results. The greatest mistake would be to enact valid policies to cope with climate change, and then abandon them because they do not produce short-term results.

Principle 4. Technology can help, but it cannot save us from ourselves. Technology is a wonderful tool. Without it, we would not be able to analyze the data we have gathered from glaciers and seabeds around the world, build the climate prediction models that help us look far into the future, and peer down from Earth's orbit at the weather. Insofar as human technology has created the industrial and post-industrial global civilization in which we now live, it is also a major factor in creating climatic instability. Some technologies may even help us to re-stabilize global climate. However, it does not follow that technology alone can provide a solution to the problem.

Principle 5. We cannot overlook steps that will improve the environment, whether they have an effect on climate or not. Certain kinds of air and water pollution such as toxic metals do not necessarily affect climate, but do affect people and other life forms. The Ice Chronicles have been used to measure improvements in air quality that are valuable to human beings, regardless of whether they make a difference in the climate. This has been a valuable intersection of science, policy, and politics.

Climate is a subset of environmental issues, and while climate is extremely important, it does not represent the whole picture of planetary habitability. Similarly, while global warming is a legitimate concern, it is not the only environmental issue that should be receiving widespread public and political attention.

Principle 6. No matter what we do, the climate will change. Whatever humans have done or will do in the millions of years they have been and will be on Earth, the climate is going to change. Moreover, it has a penchant for changing very dramatically and quickly as a consequence of the nature of the climate system itself. The reality of rapid climate change events (RCCEs) is among the most important messages to be deciphered in reading the Ice Chronicles. It is also one of the greatest benefits of having the expanded view of human history in space and time that the Ice Chronicles helps us to construct.

Principle 7. We must change with it. Human beings have become accustomed to the relatively mild and stable climate of the past several millennia, but we cannot become addicted to it. To some extent, the kinds of civilizations and societies that we consider normal are in fact abnormal, having been built during this anomalous mild period set between significantly colder periods when ice sheets covered much of the area we currently inhabit in the Northern Hemisphere. Our ancestors built nimble societies to cope with a dynamic climate system, and we need to learn from their wisdom.

Principle 8. We must continue research on climate change. While much has been learned about the climate system, much more needs to be known. For example, the influence of changes in the sun's energy output on climate is a tremendously important topic that needs much more investigation, as is the relationship between ocean circulation and climate change. Thus, while enough is known about climate change now to take action, the interaction between public policy and research should continue indefinitely.

Principle 9. We should strive more for climate predictability than control. The public debate is focused on the more obvious effects of human-induced climate change, such as rising temperatures, unstable weather patterns, and shifts in growing seasons. However, this debate needs to be expanded and extended to the issue of predictability. Years of research have now brought us to the point of understanding natural climate cycles that operate on the order of decades and longer. Yet, the predictability promised by this understanding is undermined by the "wild card" of the global human system and its impact on climate. That should be a primary concern.

Principle 10. If we cannot have global control of climate policy, we must at least have global cooperation. While "global warming" may be something of a misnomer, since the effects of climate change may not be distributed equally around the world, it will still take global efforts to influence the system to move in a more positive direction. The climate change policies do not have to be the same for every country, but the overall framework must be international in scope.

Summary

We do not submit these ten principles as the final word on climate change and environmental policy. If our work has shown us anything, it is that our assumptions about climate must change, just as climate itself changes over time. Our ten principles may well prove obsolete in the face of new information. We invite others to add to this basic foundation and build from it.

However, we do believe that making better environmental policy in the future will depend on a multifaceted understanding of the complex interactions between the global climate system and the human system that is now playing a major role in climate's behavior. If the true complexity of the challenge is fully understood, those policies will be based on scientific principles, rather than simplistic arguments pro and con. The debate will be focused on where the science shows a consensus, rather than on how to find points of conflict among researchers.

Making the right choices transcends the short-term perspectives produced by human political and economic considerations; the long-term habitability of our home planet is at stake. In the end, we return to the insights brought to us by our astronauts and cosmonauts as they took humanity's first steps into the universe: We live in a small, beautiful oasis floating through a vast and mysterious cosmos. We are the stewards of this "good Earth," and it is up to us to learn how to take good care of her.

We can look to some of the remotest regions of Earth to see what the Earth was like before the major human impacts of the last century. Places such as Antarctica provide us with an example of what natural air quality really is, although even this region is experiencing change produced by human activity, notably the Antarctic ozone hole. The air is so clean you can take a full breath and you can literally see the curvature of the Earth. What sorts of dreams for a future can be expected from this picture of the atmosphere versus the picture conjured up while flying into most industrialized cities? In which direction will human civilization evolve in the coming centuries? The answer is up to us.

Chapter References

Introduction (pp. 1–18)

Intergovernmental Panel on Climate Change Report, 1995.
Lamb, H. H. *Climate, History and the Modern World.* New York and London: Methuen & Co. Ltd, 1982, 433 pp.
McLean, William. "US, British Specialists Call on Business to Join Fight against Global Warming." *Boston Globe,* December 24, 1999, 2.
National Academy Press. *Research Strategies for the U.S. Global Change Program.* Washington, D.C.: 1990.
White, Frank. *The Overview Effect: Space Exploration and Human Evolution.* Boston: Houghton Mifflin, 1987; Reston, Va.: American Institute of Aeronautics and Astronautics, 1998.

1. Setting the Stage for Our Modern Understanding of Climate Change (pp. 19–37)

Allen, Scott. "Early Warning?" *Boston Globe,* March 6, 1995, 25, 28.
———. "Island Nations Sound Alarm on Global Warming." *Boston Globe,* February 11, 1997.
Huxley, Leonard, ed. *Scott's Last Expedition: The Journals of Captain R. F. Scott.* Boston: Beacon Press, 1957, 373.
Keeling, C. D. "Industrial Production of Carbon Dioxide from Fossil Fuels and Limestone." *Tellus* 25 (1973) 174–198.
McIntyre, A. et al. "Glacial North Atlantic 18,000 years ago: A CLIMAP Reconstruction." *Geological Society of America Memoir* 145 (1981): 43–76.
Rotty, R. M., and G. Marland. "Fossil Fuel Combustion: Recent Amounts, Patterns, and Trends of CO_2. *The Changing Carbon Cycle: A Global Analysis,* ed. J. R. Trabalka and D. E. Reichle. New York: Springer-Verlag, 1986, 484–500.
Shackleton, N. H., and N. D. Opdyke. "Oxygen Isotope and Paleomagnetic Stratigraphy of Equatorial Pacific Core V28–238: Oxygen Isotope Temperatures and Ice Volumes on a 10^2 and 10^6 Year Scale." *Journal of Quaternary Research* 3 (1991):39–55.
U.S. Department of Energy. *International Energy Annual,* DOE/EIA-0219(88). Washington, D.C.: Energy Information Administration, 1988.
White, Frank. *The Overview Effect: Space Exploration and Human Evolution.* Boston: Houghton Mifflin, 1987; Reston, Va.: American Institute of Aeronautics and Astronautics, 1998, 205–206, 261.

2. The Making of an Ice Core "Time Machine" (pp. 38–79)

Barnola, Jean-Mark, D. Raynaud, Y. S. Korotkevich, and C. Lorius. "Vostok Ice Core Provides 160,000-year Record of Atmospheric CO_2." *Nature* 329 (1987): 408–14.

Chappellaz, Jerome, J.-M. Barnola, D. Raynaud, Y. S. Korotkevich, and C. Lorius. "Atmospheric CH_4 Record over the Last Climatic Cycle Revealed by the Vostok Ice Core." *Nature* 345 (1990): 127–31.

Dibb, J., P. A. Mayewski, C. F. Buck, and S. M. Drummey. "Beta radiation from snow." *Nature* 344 (1990): 25.

Farman, J. C., B. G. Gardiner and J. D. Shanklin. "Large Losses of Total Ozone in Antarctica Reveal Seasonal CIO_x/NO_x Interaction." *Nature* 315 (1985): 207.

Huxley, Leonard, ed. *Scott's Last Expedition: The Journals of Captain R. F. Scott.* Boston: Beacon Press, 1957, 373.

"Ice Cores and Global Change," *EOS Transactions* 69 (no. 46) (1988): 1579–80.

Jouzel, Jean, C. Lorius, J. R. Petit, C. Genthon, N. I. Barkov, V. M. Kotlyakov, and V. M. Petrov. "Vostok Ice Core: A Continuous Isotope Temperature Record over the Last Climatic Cycle (160,000 years)." *Nature* 329 (1987): 403–7.

Lorius, Claude, J. Jouzel, C. Ritz, L. Merlivat, N. E. Barkov, and Y. S. Korotkevich. "150,000-year Climatic Record from Antarctic Ice." *Nature* 316 (1985): 591–95.

Mayewski, P. A. and P. A. Jeschke. "Himalayan and Trans-Himalayan Glacier Fluctuations since A.D. 1812." *Arctic and Alpine Research* 11 (3) (1979): 267–87.

Mayewski, P. A., and M. R. Legrand. "Recent Increase in Nitrate Concentration of Antarctic Snow." *Nature* 346 (1990): 258–60.

Meese, D. A., A. J. Gow, R. B. Alley, G. A. Zielinski, P. M. Grootes, M. Ram, K. C. Taylor, P. A. Mayewski, and J. F. Bolzan. "The GISP2 Depth-age Scale: Methods and Results." Journal of Geophysical Research 102 (1997): C12, 26, 411–24.

National Research Council (NRC). *Global Environmental Change: Research Pathways for the Next Decade, Overview.* National Academy of Sciences: 1999, ch. 6.

Officer, Charles, and Jake Page. *Tales of the Earth: Paroxysms and Perturbations of the Blue Planet.* New York: Oxford University Press, 1993, 180.

Petit, Jean-Robert, I. Basile, A. Leruyuet, D. Raynaud, C. Lorius, J. Jouzel, M. Stievenard, V. Y. Lipenkov, N. I. Barkov, B. B. Kudryashov, M. Davis, E. Saltzman, and V. Kotlyakov. "Four Climate Cycles in Vostok Ice Core." *Nature* 387 (1997): 359.

3. The Discovery of Rapid Climate Change Events (pp. 80–110)

Alley, R. B., D. Meese, C. A. Shuman, A. J. Gow, K. Taylor, M. Ram, E. D. Waddington, and P. A. Mayewski. "Abrupt Accumulation Increase at the Younger Dryas Termination in the GISP2 Ice Core." *Nature* 362 (1993): 527–29.

Asimov, Isaac, and Frank White. *March of the Millennia: A Key to Understanding History.* New York: Walker and Company, 1990.

Behl, R. J., and J. P. Kennett. "Brief Interstadial Events in the Santa Barbara Basin, NE Pacific, During the Past Sixty Thousand Years." *Nature* (1996) 379, 243–46.

Berger, A. "Long-term Variations of Daily Insolation and Quaternary Climatic Changes." *Journal of Atmospheric Sciences* 35, no. 12 (1978): 2362–67.

Bond, G., H. Heinrich, W. Broecker, L. Labeyrie, J. McManus, J. Andrews, S. Huon, R. Jantschik, S. Clasen, C. Simet, K. Tedesco, M. Klas, G. Bonani, and S. Ivy. "Evidence for Massive Discharges of Icebergs into the North Atlantic Ocean during the Last Glacial Period." *Nature* 360 (1992): 245–50.

Boyle, E. "Characteristics of the Deep Ocean Carbon System during the Past 150,000 Years: CO_2 Distributions, Deep Water Flow Patterns, and Abrupt Climate Change." Revelle Symposium Volume, Proceedings of the National Academy of Sciences, 1996.

Brook, L. J., T. Sowers, and J. Orchado. "Rapid Variations in Atmospheric Methane Concentration during the Past 110,000 Years." *Science* 273 (1996): 1007–91.

Chappellaz, J., T. Blunier, D. Ratnaud, J.-M. Barnola, J. Schwander, and B. Stauffer. "Synchronous Changes in Atmospheric CH_4 and Greenland Climate between 40 and 8 k yr B." *Nature* (1993): 366, 443–45.

Cuffey, K. M., G. D. Clow, R. B. Alley, M. Stuiver, E. D. Waddington, and R. W. Saltus. "Large Arctic-temperature Change at the Wisconsin-Holocene Transition." *Science* 270 (1995): 455–58.

Dansgaard, W. "Stable Isotopes in Precipitation." *Tellus* 16 (1964): 436–68.

Dansgaard, W., J. W. C. White, and S. J. Johnsen. "The Abrupt Termination of the Younger Dryas Event." *Nature* 339 (1989): 532–33.

Denton, G. H., and T. J. Hughes. "The Arctic Ice Sheet: An Outrageous Hypothesis." In *The Last Great Ice Sheets*. New York: Wiley-Interscience, 1985, Fig. 8–11, p. 460.

Denton, G. "The Influence of the Antarctic Ice Sheet on Past Global Change." *Science* (in press).

Grimm, E. C., G. L. Jacobson, Jr., W. A. Watts, B. C. S. Hansen, and K. A. Maasch. "A 50,000-year Record of Climate Oscillations from Florida and its Temporal Correlation with the Heinrich Events." *Science* 261 (1993): 198–200.

Grootes, P. M., E. J. Steig, M. Stuiver. "The Oxygen Isotope Record from Taylor Dome, Antarctica." *EOS Transactions* 76 (1994): S176.

Grootes, P. M., E. J. Steig, M. Stuiver, E. D. Waddington, D. L. Morse. "GISP2-Taylor Dome Oxygen Isotope Ratios." *Quaternary Research* (in press).

Grootes, P. M., M. Stuiver. "Oxygen 18/16 Variability in Greenland Snow and Ice with 10^3 to 10^5-year time resolution." *Journal of Geophysical Research* 102 (1997): 26455–70.

Grootes, P. M., M. Stuiver, J. W. C. White, S. J. Johnsen, and J. Jouzel. "Comparison of Oxygen Isotope Records from the GISP2 and GRIP Greenland Ice Cores." *Nature* 366 (1993): 552–54.

Hays, J. D., J. Imbrie, and N. J. Shackleton. "Variations in the Earth's Orbit: Pacemaker of the Ice Ages." *Science* 194 (1976): 1121–32.

Hodge, S. M., D. L. Wright, J. A. Bradley, R. W. Jacobel, N. Skou, and B. Vaughn. "Determination of the Surface and Bed Topography in Central Greenland." *Journal of Glaciology* 36, no. 122 (1990): 17–30.

Hood, L. L., and J. L. Jirikowic. "Recurring Variations of Probable Solar Origin in the Atmospheric C^{14} Time Record." *Geophysical Research Letters* 17, no. 1 (1990): 85–88.

Imbrie, J., and K. Imbrie. *Ice Ages, Solving the Mystery*. London: MacMillan Press, 1979.

Johnson, S. J., H. B. Clausen, W. Dansgaard, K. Fuhrer, N. Gundestrup, C. V. Hammer, P. Iverson, J. Jouzel, B. Stauffer, and J. P. Steffensen. "Irregular Glacial Interstadials Recorded in a New Greenland Ice Core." *Nature* 359 (1992): 311–13.

Jouzel, J., C. Lorius, J. R. Petit, C. Genthon, N. I. Barkov, V. M. Kotttlyakov, and V. M. Petrov. "Vostok Ice Core: A Continuous Isotope Temperature Record Over the Last Climatic Cycle (160,000 Years)." *Nature* 329 (1987): 402–8.

Legrand, M., and P. A. Mayewski. "Glaciochemistry of Polar Ice Cores: A Review." *Reviews of Geophysics* 35 (1997): 219–43.

Lorius, C., and L. Merlivat. "Distribution of Mean Surface Stable Isotope Values in East Antarctica: Observed Changes with Depth in a Coastal Area." In *Isotopes and Impurities in Snow*. IAHS Publ. 118 (1977): 125–37.

Lorius, C., J. Jouzel, C. Ritz, L. Merlivat, N. I. Barkov, Y. S. Korotkevitch, and V. M. Kotlyakov. "A 150,000-year Climatic Record from Antarctic Ice." *Nature* 316 (1985): 591–96.

Mayewski, P. A., G. H. Denton, and T. Hughes. "Late Wisconsin Ice Margins of North America." In *The Last Great Ice Sheets*, eds. G. Denton and T. Hughes. New York: Interscience Publ., John Wiley and Sons, 1981.

Mayewski, P. A., L. D. Meeker, M. S. Twickler, S. I. Whitlow, Q. Yang, W. B. Lyons, and M. Prentice. "Major Features and Forcing of High Latitude Northern Hemisphere Atmospheric Circulation over the Last 110,000 Years." *Journal of Geophysical Research* 102 (1997): C12, 26, 345–26, 366.

Mayewski, P. A., L. D. Meeker, S. Whitlow, M. S. Twickler, M. C. Morrison, R. B. Alley, P. Bloomfield, and K. Taylor. "The Atmosphere during the Younger Dryas." *Science* 261 (1993): 195–97.

Meeker, L. D., P. A. Mayewski, M. S. Twickler, S. I. Whitlow, and D. A. Meese. "A 110,000-year History of Change in Continental Biogenic Emissions and Related Atmospheric Circulation Inferred from the Greenland Ice Sheet Project Ice Core." *Journal of Geophysical Research* 102 (1997): C12, 26, 489–26, 504.

Meese, D. A., R. B. Alley, A. J. Gow, P. Grootes, P. A. Mayewski, M. Ram, K. C. Taylor, E. D. Waddington, and G. Zielinski. "The Accumulation Record from the GISP2 Core as an Indicator of Climate Change throughout the Holocene." *Science* 266 (1994): 1680–82.

National Research Council (NRC). *Global Environmental Change: Research Pathways for the Next Decade, Overview.* Washington, D.C.: National Academy of Sciences, 1998, ch. 6.

Oeschger, H. "Nuclear and Chemical Dating Techniques: Interpreting the Environmental Record." In *American Chemical Society Symposium Series*, No. 176, ed. L. A. Currie, 1982, 5–42.

Porter, S. C., and A. An. "Correlation between Climate Events in the North Atlantic and China during the Last Glaciation." *Nature* 375 (1995): 305–8.

Sachs, J., and S. J. Lehman. "Subtropical Atlantic Temperatures 60,000–30,000 Years Ago." *Science* 286 (1999): 756–89.

Saltzman, B., A. Sutera, and A. Evenson. "Structural Stochastic Stability of a Simple Auto-oscillatory Climate Feedback System." *Journal of Atmospheric Research* 38, no. 3 (1981): 494–503.

Shackleton, N. J., and N. D. Opdyke. "Oxygen Isotope and Paleomagnetic Stratigraphy of Pacific Core V28–239 Late Pliocene to Latest Pleistocene." Geological Society of America, *Memoir* 145 (1976): 449–64.

Sonett, C., and S. A. Finney. "The Spectrum of Radiocarbon." *Philosophical Transactions of the Royal Society of London* A 330 (1990): 413–26.

Stuiver, M., T. F. Braziunas, B. Becker, and B. Kromer. "Climatic, Solar, Oceanic, and Geomagnetic Influences on Late-glacial and Holocene Atmospheric C^{14}/C^{12} Change." *Quaternary Research* 35, no. 1 (1991): 1–24.

Stuiver, M., P. M. Grootes, and T. F. Braziunas. "The GISP2 18O Climate Record of the Past 16,500 Years and the role of the Sun, Ocean and volcanoes." Quaternary Research 44 (1995): 341–54.

Suess, H. E. "The Radiocarbon Record in Tree Rings of the Last 8,000 Years." *Radiocarbon* 22 (1980): 200–209.

Taylor, K. C., P. A. Mayewski, M. S. Twickler, S. I. Whitlow. "Biomass Burning Recorded in the GISP2 Ice Core: A Record from Eastern Canada." *The Holocene* 6, no. 1 (1996): 1–6.

Whitlow, S. I., P. A. Mayewski, G. Holdsworth, M. S. Twickler, and J. Dibb. "An Ice Core-Based Record of Biomass Burning in North America." *Tellus* 46B (1994): 234–42.

Yang, Q., P. A. Mayewski, S. I. Whitlow, and M. S. Twickler. "Major Features of Geochemistry over the Last 110,000 Years in the GISP2 Ice Core." *Journal of Geophysical Research* 102, no. 22: (1997): 289–99.

4. Climate Change and the Rise and Fall of Civilizations (pp. 111–125)

Alley, R. B., P. A. Mayewski, T. Sowers, M. Stuiver, K. C. Taylor, and P. U. Clark. "Holocene Climate Instability: A Large Event 8000–8400 Years Ago." *Geology* 25 (6) (1997): 402–06.

Alley, R. B., D. Meese, C. A. Shuman, A. J. Gow, K. Taylor, M. Ram, E. D. Waddington, and P. A. Mayewski. "Abrupt Increase in Greenland Snow Accumulation at the End of the Younger Dryas Event." *Nature* 362 (1993): 527–29.

Andrews, J. T., and J. Ives. "Late and Postglacial Events (<10,000 B.P.) in Eastern Canadian Arctic with Particular Reference to the Cockburn Moraines and the Breakup of the Laurentide Ice Sheet." In *Climate Changes in the Arctic Areas during the Last 10,000 Years*, ed. Y. Vasari, H. Hyrarinen, and S. Hicks. Oulu, Finland: University of Oulu, 1972, 149–76.

Bar-Yosef, O. "The Role of Climate in the Interpretation of Human Movements and Cultural Transformations in Western Asia." In *Paleoclimate and Evolution with Emphasis on Human Origins*, ed. E. S. Vrba, G. H. Denton, T. C. Partridge, and L. H. Buckle. New Haven and London: Yale University Press, 1995, 507–23.

Bond, G., W. Showers, M. Cheseby, R. Lotti, P. Almasi, P. deMenocal, P. Priore, H. Cullen, I. Hajdas, and G. Bonani. "A Pervasive Millennial-scale Cycle in North Atlantic Holocene and Glacial Times." *Science* 278 (1997): 1257–66.

Cullen, H. M., P. M. deMenocal, S. Hemming, G. Hemming, F. H. Brown, T. Guildrson, and F. Sirocko. "Climate Change and the Collapse of the Akkadian Empire: Evidence from the Deep Sea." *Geology* 28, no. 4 (2000): 379–82.

Denton, G. H., and W. Karlen. "Holocene Climatic Variations: Their Pattern and Possible Cause." *Quaternary Research* 3 (1973): 155–205.

Harvey, L. D. D. "Solar Variability as a Contributing Factor to Holocene Climate Change." *Progress Physical Geography* 4 (1980): 487–530.

Hodell, D., J. Curtis, and M. Brenner. "Possible Role of Climate in the Collapse of Classic Maya Civilization." *Nature* 375 (1995): 391–94.

Linden, Eugene. "Warnings from the Ice." *Time* magazine, April 14, 1997.

Mayewski, P. A., L. D. Meeker, S. Whitlow, M. S. Twickler, M. C. Morrison, R. B. Alley, P. Bloomfield, and K. Taylor. "The Atmosphere during the Younger Dryas." *Science* 261 (1973): 195–97.

O'Brien, S. R., P. A. Mayewski, L. D. Meeker, D. A. Meese, M. S. Twickler, and S. I. Whitlow. "Complexity of Holocene Climate as Reconstructed from a Greenland Ice Core." *Science* 270 (1995): 1962–64.

Stager, J. C., and P. A. Mayewski. "Abrupt Mid-Holocene Climatic Transitions Registered at the Equator and the Poles." *Science* 276 (1997): 1834–36.

Taylor, K. C., C. U. Hammer, R. B. Alley, H. B. Clausen, D. Dahl-Jensen, A. J. Gow, N. S. Gundestrup, J. Kipfstuhl, J. C. Moore, and E. D. Waddington. "Electrical Conductivity Measurements from the GISP2 and GRIP Greenland Ice Cores." *Nature* 366 (1993): 549–52.

Weiss, H., M.-A. Courty, W. Wetterstrom, F. Guichard, L. Senior, R. Meadow, and A. Curnow. "The Genesis and Collapse of Third Millennium North Mesopotamian Civilization." *Science* 261 (1993): 995–1004.
Zielinski, G. A., P. A. Mayewski, L. D. Meeker, K. Gronvald, M. S. Germani, S. I. Whitlow, M. S. Twickler, and K. Taylor. "Volcanic Aerosol Records and Tephrochronology of the Summit, Greenland, Ice Cores." *Journal of Geophysical Research* 102 (1997); C12, 26, 625–26, 640.

5. The Last Thousand Years of Climate Change (pp. 126–160)

Broad, William J. "Another Possible Climate Culprit: The Sun." New York *Times*, September 23, 1997, C1.
Etheridge, D. H. "Natural and Anthropogenic Changes in Atmospheric Carbon Dioxide and Methane over the Last 1,000 Years." Ph.D. dissertation, School of Earth Sciences, University of Melbourne, 1999.
Etheridge, E. M., L. P. Steele, R. L. Langenfelds, R. J. Francey, J.-M. Barnola, J.-M. and V. I. Morgan. "Natural and Anthropgenic Changes in Atmospheric CO2 over the last 1,000 Years from Air in Antarctic Firn and Ice." *Journal of Geophysical Research* 101, no. D2 (1996): 4115–28.
Grove, J. M. *The Little Ice Age.* London: Methuen, 1988.
Hansen, J., and S. Lebedeff. "Global Surface Air Temperatures: Update through 1987." *Geophysical Research Letters* 15 (1988) 323–26.
Hayden, Thomas, and Sharon Begley. "Cold Comfort." *Newsweek,* August 11, 1997.
Khalil, M. A. K., and R. A. Rasmussen. "Sources, Sinks, and Seasonal Cycles of Atmospheric Methane." *Journal of Geophysical Research* 88 (1988): 5131–41.
Khalil, M. A. K. and R. A. Rasmussen. "Nitrous Oxide: Trends and Global Mass Balance over the Last 300 Years." *Annals of Glaciology* 10 (1988): 73–79.
Mann, M. E., R. S. Bradley, and M. K. Hughes, "Global Scale Temperature Patterns and Climate Forcing over the Past Six Centuries." *Nature* 392 (1998) 779–87.
Mayewski, P. A., G. Holdsworth, M. J. Spencer, S. Whitlow, M. S. Twickler, M. C. Morrison, K. F. Ferland, and L. D. Meeker. "Ice Core Sulfate from Three Northern-Hemisphere Sites: Source and Temperature Forcing Implications." *Atmospheric Environment* 27a, nos. 17/18 (1993b): 2915–19.
Mayewski, P. A., L. D. Meeker, M. C. Morrison, M. S. Twickler, S. Whitlow, K. K. Ferland, D. A. Meese, M. R. Legrand, and J. P. Steffenson, "Greenland Ice Core 'Signal' Characteristics: An Expanded View of Climate Change," *Journal of Geophysical Research* 98, no. D7 (1993a): 12,839–47.
Mayewski, P. A. and P. A. Jeschke. "Himalayan and Trans-Himalayan Glacier Fluctuations since A.D. 1812." *Arctic and Alpine Research* 11, no. 3 (1979): 267–87.
Mayewski, P. A., W. B. Lyons, M. J. Spencer, M. S. Twickler, C. F. Buck, and S. Whitlow. "An Ice Core Record of Atmospheric Response to Anthropogenic Sulphate and Nitrate." *Nature* 346, no. 6284 (1990): 554–56.
Mayewski, P. A., W. B. Lyons, M. J. Spencer, M. S. Twickler, P. B. Koci, P. Dansgaard, C. Davidson, and R. Honrath. "Sulfate and Nitrate Concentrations from a South Greenland Ice Core." *Science* 232 (1986): 975–77.
Meeker, L. D., and P. A. Mayewski. "A 1,400-year-long Record of Atmospheric Circulation over the North Atlantic and Asia." *The Holocene* (in press).
Mergen, Bernard. *Snow in America.* Washington and London: Smithsonian Institution Press, 1997, 1.

O'Brien, S. R., P. A. Mayewski, L. D. Meeker, D. A. Meese, M. S. Twickler, and S. I. Whitlow. "Complexity of Holocene Climate as Reconstructed from a Greenland Ice Core." *Science* 270 (1995): 1962–64.

Officer, Charles, and Page, Jake. *Tales of the Earth: Paroxysms and Perturbations of the Blue Planet.* New York: Oxford University Press, 1993, 97.

Pearman, G. I., D. Etheridge, F. deSilva, and P. J. Fraser. "Evidence of Changing Concentrations of Atmospheric CO_2, N_2O, and CH_4 from Air Bubbles in Antarctic Ice." *Nature* 320 (1986): 248–50.

Stuiver, M., and T. F. Braziunas. "Sun, Ocean, Climate and Atmospheric $^{14}CO_2$: An Evaluation of Causal and Spectral Relationships." *The Holocene* 3 (1993) 289–305.

Suplee, Curt. "It's Not Easy to Chart the Earth's Temperature." Washington *Post,* reprinted in *Valley News,* February 28, 2000, C1.

Zardini, D., D. Raynaud, D. Scharffe, and W. Seiler. "N_2O Measurements of Air Extracted from Antarctic Ice Cores: Implications on Atmospheric N_2O Back to the Last Glacial-Interglacial Transition." *Journal of Atmospheric Chemistry* 8 (1989): 189–201.

6. Climate Change: The Real Impact (pp. 161–178)

Anderson, R. "Long-term Changes in the Frequency of Occurrence of El Niño Events." In *El Niño: Historical and Paleoclimate Aspects of the Southern Oscillation,* ed. H. F. Diaz and V. Markgraf, Cambridge: Cambridge University Press, 1992, 193–200.

Barry R. G., and R. J. Chorley. *Atmosphere, Weather, and Climate,* 6th ed., London and New York: Routledge Press, 1992, 233.

Chandler, David L. "Alaska Is Feeling the Heat," *Boston Globe,* September 15, 1997, C1.

Cotton William R., and Roger A. Pielke. *Human Impacts on Weather and Climate.* 6th ed. Cambridge: Cambridge University Press, 1995.

Crutzen, J., and. J. W. Birks. "The Atmosphere after a Nuclear War: Twilight at Noon." *Ambio* 11 (1981): 114–25.

Detwyler, T. R., ed. *Man's Impact on the Climate.* New York: McGraw Hill, 1971.

Easterling, D. R., J. L. Evans, P. Groisman, P. Ya, T. R. Karl, K. E. Kunkel, and P. Ambenje. "Observed Variability and Trends in Extreme Climate Events: A Brief Review." *Bulletin of the American Meteorological Society* 81, no. 3 (2000): 417–25.

Fialka, John J. "U.S. Study on Global Warming May Overplay Dire Side." *Wall Street Journal,* Friday, May 26, 2000, A24.

Jones, P. D. "Hemispheric Surface Air Temperature Variations: A Reanalysis and Update to 1993." *Journal of Climate* 7 (1994): 1794–1802.

Karl, T. R., and D. R. Easterling. "Climate Extremes: Selected Review and Future Research Directions." *Climate Change* 42 (1999): 309–25.

Manabe S., and R. Stouffer. "Simulation of Abrupt Climate Change Induced by Freshwater Input to the North Atlantic Ocean." *Nature* 378 (1995): 165–67.

McElroy, Michael B. "A Warming World." *Harvard Magazine,* November–December 1997, 35–37.

Parker, D. E., P. D. Jones, A. Bevan, and C. K. Folland. "Interdecadal Changes of Surface Temperature Since the Late 19th Century." *Journal of Geophysical Research* 99 (1994): 14,373–99.

Zielinski, G. A., P. A. Mayewski, L. D. Meeker, S. Whitlow, and M. S. Twickler. "Potential Atmospheric Impact of the Toba Mega-Eruption ~71,000 Years Ago." *Geophysical Research Letters* 23, no. 8 (1996): 837–40.

7. Confronting the Choices (pp. 179–200)

Chandler, David L. "Issue Puts Scientific Ethics to Test." *Boston Globe,* December 10, 1997a, A19.
———. "Low-tech Ways to Reverse the Buildup." *Boston Globe,* December 8, 1997b, D1.
———. Specialists Differ on Plan's Impact." *Boston Globe,* December 12, 1997c, A26.
Global Change, "Heat Waves Take Heavy Toll on Urban Poor." February, 1996, 3.
Colwell, Rita R. "Global Climate and Infectious Disease: The Cholera Paradigm." *Science* 274, December 20, 1996, 2025–31.
Hahn, Robert W., and. Robert N. Stavins. *What Has the Kyoto Protocol Wrought? The Real Architecture of International Tradable Permit Markets.* Washington, D.C.: The AEI Press, 1999.
Howe, Peter J. "Rich, Poorer Nations Set to Clash in Kyoto." *Boston Globe,* December 1, 1997, 1.
Intergovernmental Panel on Climate Change. *Climate Change 1995. The Science of Climate Change.* Contribution of Working Group 1 to the Second Assessment Report of the IPCC. Cambridge: Cambridge University Press, 1995.
Lakshmanan, Indira A. R. "Accord Set on Cutting Emissions." *Boston Globe,* December 11, 1997, A30.
Lamb, H. H. *Climate, History and the Modern World.* New York and London: Methuen & Co, Ltd., 1982.
Mann, Charles. "The Weather Tamers." *Science Digest,* November 1983, 70.
Mayewski, P. A. and J. W. Attig. "A Recent Decline in Available Moisture in Northern Victoria Land, Antarctica." *Journal of Glaciology* 20, no. 84 (1978): 593–94.
Mayewski, P. A., J. W. Attig, and D. J. Drewry. "Pattern of Ice Surface Lowering for the Rennick Glacier, Northern Victoria Land, Antarctica. *Journal of Glaciology* 22, no. 86 (1979): 53–65.
McDougall, Walter. *The Heavens and the Earth: A Political History of the Space Age.* Baltimore, Md.: Johns Hopkins University Press, 1985, 1997.
McElroy, Michael B. "A Warming World." *Harvard Magazine,* November–December 1997, 35–37.
McGrory, Brian. "Negotiators on Warming Get Big Chill." *Boston Globe,* December 12, 1997, A26.
Warrick, Joby. "Earth Summit Opens with Rift over U.S. Air Pollution Curbs." *Boston Globe,* June 24, 1997, A10.

8. Learning to Live in a Changing World (pp. 201–214)

Chandler, David L., "Specialists Differ on Plan's Impact." *Boston Globe,* December 12, 1997, A26.
Gore, Al. *Earth in the Balance.* New York: Plume/Penguin, 1992, 368.
Stevens, William K. "Doubts on Cost Are Bedeviling Climate Policy." New York *Times,* December 6, 1997, A1.

Bibliography

American Geophysical Union. Greenland Summit Ice Cores—Greenland Ice Sheet Project 2 and Greenland Ice Core Project. *Journal of Geophysical Research* (1997).

Asimov, Isaac, and Frank White. *March of the Millennia: A Key to Understanding History.* New York: Walker and Company, 1990.

Barry, R. G., and R. J. Chorley. *Atmosphere, Weather, and Climate.* 6th ed. London and New York: Routledge Press, 1992.

Bradley, Raymond S. *Paleoclimatology: Reconstructing Climates of the Quaternary.* 2nd ed. New York: Harcourt, Academic Press, 1999.

Cotton, William R., and Roger A. Pielke. *Human Impacts on Weather and Climate.* Cambridge: Cambridge University Press, 1995.

Detwyler, T. R., ed. *Man's Impact on the Climate.* New York: McGraw Hill, 1971.

Gore, Al. *Earth in the Balance.* New York: Plume/Penguin, 1992.

Huxley, Leonard, ed. *Scott's Last Expedition: The Journals of Captain R. F. Scott.* Boston: Beacon Press, 1957.

Lamb, H. H. *Climate, History and the Modern World.* New York and London: Methuen & Co. Ltd, 1982.

McDougall, Walter. *The Heavens and the Earth: A Political History of the Space Age.* Baltimore, Md.: Johns Hopkins University Press, 1985, 1997.

Mergen, Bernard. *Snow in America.* Washington and London: Smithsonian Institution Press, 1997.

National Academy Press. *Research Strategies for the U.S. Global Change Program.* Washington, D.C.: National Academy Press, 1990.

National Research Council (NRC). *Global Environmental Change: Research Pathways for the Next Decade, Overview.* Washington, D.C.: National Academy of Sciences, 1998.

Officer, Charles, and Jake Page. *Tales of the Earth: Paroxysms and Perturbations of the Blue Planet.* New York: Oxford University Press, 1993.

White, Frank. *The Overview Effect: Space Exploration and Human Evolution.* Boston: Houghton Mifflin, 1987; Reston, Va.: American Institute of Aeronautics and Astronautics, 1998.

Wilson, R. C. L., S. A. Drury, and J. L. Chapman. *The Great Ice Age—Climate Change and Life.* London and New York: Routledge Press, 2000.

Index